POLLUTION CHARACTERIZATION AND QUANTIFICATION IN THE AGRICULTURE SECTOR

APRIL 2024

ADB

ASIAN DEVELOPMENT BANK

© 2024 Asian Development Bank
6 ADB Avenue, Mandaluyong City, 1550 Metro Manila, Philippines
Tel +63 2 8632 4444; Fax +63 2 8636 2444
www.adb.org

Some rights reserved. Published in 2024.

ISBN 978-92-9270-670-8 (print); 978-92-9270-671-5 (PDF); 978-92-9270-672-2 (ebook)
Publication Stock No. TCS240236
DOI: http://dx.doi.org/10.22617/TCS240236

Corrigenda to ADB publications may be found at http://www.adb.org/publications/corrigenda.

Notes:
In this publication, "$" refers to United States dollars.
ADB recognizes "China" as the People's Republic of China and "Vietnam" as Viet Nam.

On the cover: Aquaculture in Java, Indonesia; fruit processing in the Philippines; and small cattle farm in a rural area in the Lao People's Democratic Republic (photos by Xin Ren).

Contents

Tables, Figures, and Boxes   v

Acknowledgments   vi

Abbreviations   vii

Executive Summary   viii

I. Introduction   1

The Need for and Application of Norms and Reference Values   1
Objectives and Coverage of the Study   5

II. Pollution Characteristics of Animal Husbandry   7

Water Use and Wastewater Generation in Animal Husbandry   7
Characteristics of Animal Husbandry Wastewater   9
Solid Waste Characteristics in Animal Husbandry   10

III. Pollution Characteristics of Aquaculture   12

Water Use and Wastewater Generation in Aquaculture   12
Characteristics of Aquaculture Effluents   13
Solid Waste Characteristics in Aquaculture   15

IV. Pollution Characteristics of Meat Processing   16

Water Use and Wastewater Generation in Meat Processing   16
Characteristics of Slaughtering Wastewater   19
Solid Waste Characteristics in Slaughtering   20

V. Pollution Characteristics of Fish Processing   22

Water Use and Wastewater Generation in Fish Processing   22
Characteristics of Fish Processing Wastewater   22
Solid Waste Characteristics in Fish Processing   27

VI. Pollution Characteristics of Fruit and Vegetable Processing   **28**

Water Use and Wastewater Generation in Fruit and Vegetable Processing   28
Solid Waste Characteristics in Fruit and Vegetable Processing   31

VII. Discussion and Conclusions   **32**

References   **36**

Tables, Figures, and Boxes

Tables

1	Water Use and Wastewater Estimate in a Livestock Farm	3
2	Estimate of Slaughterhouse Wastewater Treatment	5
3	Wastewater from Livestock Holding	8
4	Wastewater Characteristics of Cattle Holding	10
5	Wastewater Characteristics of Pig Holding	10
6	Unit Waste Generation from Livestock Holding	11
7	Unit Water Use for General Purposes and Associated Wastewater	12
8	Effluent Characteristics of Aquaculture (Flow-Through System)	13
9	Comparison of Aquaculture Effluents with Municipal Wastewater	14
10	Water Norms for Slaughterhouse	17
12	Water Consumption in Poultry Slaughtering	18
11	Average Water Use during Main Steps in Slaughtering	18
13	Wastewater Characteristics from Livestock Slaughtering	19
14	Wastewater Characteristics from Poultry Slaughtering	20
15	Unit Generation of Solid Waste from Slaughtering	21
16	Wastewater Characteristics of Fish Processing by Source	24
17	Wastewater Characteristics of Fish Processing by Species	25
18	Wastewater Characteristics by Main Fish Production Process and Step	26
19	Wastewater Characteristics of Plant-Based Processing	29
20	Wastewater by Major Processing Types of Fruit and Vegetable	30
21	Characteristics of Wastes from Major Tropical Fruit Processing	31

Figures

1	Daily Manure Production of Different Animals and Humans	15
2	Water Consumption vs. Slaughterhouse Capacity	17

Boxes

1	Determining Pollution Characteristics	2
2	Can This Technology Treat Wastewater into Compliance?	3
3	The Myth of Computer Models vs. Other Methods to Predict Impacts	34

Acknowledgments

This study was initiated and led by Xin Ren, senior environmental specialist of the Asian Development Bank (ADB), funded under a regional technical assistance (TA) on Strengthening Safeguards Management in Southeast Asia (TA 9647). She authored Chapters 1, 3, 5, 6, and 7 and contributed to and finalized the rest. Ernesto Dela Cruz, senior consultant and environmental–chemical engineer of more than 30 years, drafted Chapters 2 and 4, and contributed to Chapter 6. Adam Yazid, environmental consultant, contributed to Chapter 5 in compiling data from literature research. This study draws upon the authors' collective experience and insights into different facets of pollution.

Xin Ren has worked closely on rural projects in Southeast Asian countries, focusing on environmental management. Before joining ADB, she worked at the World Bank on diverse projects on urban development, energy, and transport to rural sectors. Apart from her 15 years in multilateral development banks, she also worked for years in the United Nations Framework Convention on Climate Change secretariat; the United Nations Environmental Program on waste management; and in the People's Republic of China on pollution control, hazardous waste management, and cleaner production.

The report benefited from valuable comments and suggestions from peer reviewers: Beatrice Yulo Gomez, principal environmental specialist; Ka Seen Gabrielle Chan, environmental specialist; and Melissa Moyano Manguiat, senior safeguard officer; all from private sector operation of the ADB Office of Safeguards, and Nurlan Djenchuraev, senior environmental specialist, region 1 of Office of Safeguards. The study could not have been realized without the support of Antoine Morel, coordinator for environmental work in the ADB Southeast Asia Department and TA 9647; and Nianshan Zhang, head of the ADB Office of Safeguards. Special thanks go to the ADB Library for the literature survey, especially Loureal Camille Inocencio, for the excellent service. The authors remain responsible for all residual shortcomings.

Abbreviations

ADB	Asia Development Bank
ASAE	America Society of Agricultural Engineers
BOD	biochemical oxygen demand
CFU	Colony Forming unit
COD	chemical oxygen demand
DEWATS	decentralized wastewater treatment technology
DMC	developing member country
EHS	environmental, health, and safety
EIA	environmental impact assessment
FAO	Food and Agriculture Organization
IFC	International Finance Corporation
MEE	Ministry of Ecology and Environment (People's Republic of China)
NH3-N	ammonia nitrogen
NH4-N	ammonium
NO3-N	nitrate nitrogen
O&G	oil and grease
PAES	Philippines Agriculture Engineering Standard
PRC	People's Republic of China
RAS	recirculating aquaculture system
SMEs	small and medium-sized enterprises
TKN	total Kjeldahl nitrogen
TSS	total suspended solids
UNEP	United Nation Environment Program
US	United States
USEPA	United States Environmental Protection Agency
WHO	World Health Organization

Executive Summary

Projects of the Asian Development Bank (ADB) in the agriculture sector mainly include animal production and processing, aquaculture and aquatic products processing, and plant-based processing subsectors. These subsectors deal with different materials, products, production processes, and operational practices. As a result, the pollution they generate are equally complex, and in some sectors, severe.

To tackle the various types of pollution, their characteristics need to be assessed and quantified, especially for point sources, according to the standard practice of environmental impact assessment in many countries. These characteristics include wastewater volume, major pollutants, concentration in terms of typical parameters regulated by discharge standards of countries, as well as nature and respective amounts of solid waste.

Characterizing and quantifying pollution makes it possible to evaluate (i) if they can meet the applicable environmental standards; (ii) if the control and treatment technology proposed can reduce the pollution into compliance with applicable standards; and (iii) if not, what are the alternative or additional remedies. Examples in Chapter 1 well illustrate these tasks and challenges facing project assessment.

Pollution is quantified basically by production scale multiplied by the unit discharge of wastes (liquid or solid). These are called reference values or industrial norms. The International Finance Corporation's Environmental, Health, and Safety Guidelines provide unit water use or solid waste for some sectors, but not for aquaculture, livestock husbandry, and plant-based processing (e.g., fruit). They lack data on wastewater volume and their characteristics for all agriculture subsectors.

Local conditions do not always allow for sample tests to obtain first-hand data, which should also be cross-checked by literature survey, and both are again constrained by time pressure at the project level. Even if both are obtainable, project designers and assessors still need to judge which values are more reasonable for their project. All these point to the need for a repository of refence values, hence this study to facilitate pollution quantification and prediction.

Using desk reviews and practical project experiences, the study illustrates the tasks and challenges facing project assessment (Chapter 1) and demonstrates step-by-step how to apply the data in standard impact assessment. Agriculture subsectors are diverse, resulting in wide variation in unit water use, pollution discharge, and concentration range, except in the livestock subsector. Such variations have not only rendered this study more challenging than expected but also constrain their use in pollution estimation (Chapters 2–6).

Therefore, case-by-case testing of pollution and wastes at the project cannot be substituted entirely and should always be preferred.

The study further discussed the standard methods, and the related confusion in estimating pollution in terms of pollution strength and quantity at discharge point before and after pollution treatment (Chapter 7). These are not only needed for checking compliance with discharge standards, but also lay the basis for predicting concentration at the receiving end after dispersing through environmental media like air and water, and the final result such as ambient environmental quality. Finally, the pros and cons of computerized modeling for such prediction versus analogy and extrapolation are discussed.

ADB's safeguard policies aim to avoid and minimize adverse environmental and social consequences of its projects and operations. Such objective can only be realized through the project design (from feasibility study to detailed design) and execution (i.e., construction and operation). Each of these steps on the technical side can derail the project, no matter how good the environmental impact assessment and their action plans are. Tightening key steps on the technical side therefore can better foster mainstreaming environmental considerations in project design and execution than merely strengthening impact assessment.

I. Introduction

The Need for and Application of Norms and Reference Values

Agricultural activities are diverse, covering horticulture and plant-based processing, animal husbandry and processing, aquaculture and fish products processing, and agricultural waste utilization and by-products production. They deal with vastly different raw materials, products, production processes and technologies, as well as operational practices. Many of them, especially those related to animals, generate pollution that are equally diverse and complex, with potentially adverse impacts and risks on water, air, soil, and ecosystems.

In tackling the potential environmental, health, and safety (EHS) impacts of the agriculture sector, first and foremost, the characteristics of pollution, both in nature and scale, need to be understood. For water pollution, this means wastewater volume and major pollutants in the form of typical parameters like biochemical oxygen demand (BOD) and their concentrations. For solid waste, characteristics relate to their nature (e.g., organic, hazardous or not) and amount. All these are determined at the feasibility stage based on raw materials used, products generated, and production methods or practices used.

Without knowledge of pollution characteristics, it is impossible to quantify pollution in the environmental impact assessment (EIA) and evaluate if the pollution can meet the applicable discharge standards or not. The technical design developed at the feasibility study normally includes pollution control and treatment technologies. However, feasibility studies do not always justify or assess whether such technology can bring down the pollution into compliance. Nor are other pollution control options that should be integral to modern facilities considered.

These are mainly because experts on technical design and feasibility studies are specialized in the sector or industry, not necessarily on the environment. In many developing countries, there are guidelines on feasibility studies but not by subsector, for example, animal husbandry. Some have general guidelines on EIA by major types of impacts—pollution or ecological, as in the People's Republic of China (PRC)—but not by sector.

Feasibility studies basically justify if a project proposed is technically and financially feasible or not, and develop the technical design in the process. Over time, many countries require it to assess if a project is also acceptable on the socioenvironmental front. With the latter assessment becoming highly specialized that sector experts can hardly cover everything already, EIA has become stand-alone. The final decision on investment is made based on the results of feasibility studies and EIA, according to regulations in many countries nowadays.

There are many EHS guidelines, notably those of the International Financial Corporation (IFC) and the World Bank (WB), which cover many sectors and subsectors. These include industry norms on water use or solid waste generation per unit for some agriculture subsectors, except aquaculture, livestock husbandry, and plant-based processing (e.g., fruit). Even more lacking are data and information on pollution volume and their characteristics and concentration.

The challenges facing EIA especially in agriculture subsectors or industry projects are illustrated in the examples in Boxes 1 and 2. Taking the livestock farm example in Box 1, the first step to quantify solid waste is to find out the most relevant industrial norms or pollution coefficients, e.g., manure amount per day per head for different animals under various conditions (e.g., penned, free-range, or mixed). Given the discrepancy among different sources, this requires judgment calls by professionals. In this example, the EHS guideline for mammalian livestock production is equivalent to 24 kilograms (kg) per day per head of small caw (close to reality) and 1.4 kg per day per full-grown pig. When multiplying by the respective number of animals, the total manure generation is about 28 tons per day, close to estimates in the feasibility study.

Box 1: Determining Pollution Characteristics

A proposed livestock quarantine farm is designed with a holding capacity of 1,000 cattle and 2,000 pigs. In the draft feasibility study, livestock experts estimated that total manure generated will be about 26 tons per day, and urine will be roughly of an equal amount. However, there is no estimate of wastewater amount and its characteristics. Standard practice entails that the environmental impact assessment should fill the gap and verify if the feasibility study estimates on pollution are reasonable or not.

Source: Author.

The EIA preparers cross-checked other sources for waste norms and found that in the PRC, manure generation ranges from 10 to 20 kg/day/caw and about 2 kg/day/pig, close to those in the EHS guidelines. This example shows that feasibility study and EIA preparers constantly need to judge and justify which numeric norms are more suitable to use. It also highlights the need to compile data on unit discharge from different sources for cross-checking and as repository for easier use.

Box 2 illustrates wastewater, another major pollution source from animal husbandry that contributes to this subsector as the top polluter in rural areas of many countries. The EIA needs to quantify wastewater amount and characterize its major pollutants and their typical concentration range. Yet, the above IFC EHS guidelines for the subsector do not have such information.

> ### Box 2: Can This Technology Treat Wastewater into Compliance?
>
> A new slaughterhouse is designed with the capacity of slaughtering 100 pigs per day. A decentralized wastewater treatment technology system (DEWATS) is proposed in the feasibility study mainly because the DEWATS provider, a German bilateral agency, has been active in the region. Therefore, DEWATS is more readily available for small-scale wastewater treatment. However, there is no information on slaughterhouse wastewater in the feasibility study, either by volume or concentration, let alone justification if DEWATS can treat it to meet the discharge standard of the country. Such tasks fall on the environmental impact assessment as per standard practice in most countries.
>
> Source: Author.

Therefore, EIA preparers in Box 2 turned to the literature and other projects' EIAs to obtain the unit discharge and their typical concentration range for key pollution parameters regulated in most countries. The results on wastewater volume are summarized in Table 1 with further detailed explanation in Chapter 2. Industry practice is to use water consumption data to estimate and verify wastewater volume, as the former is easier to obtain and more accurate. The conversion rate for processes without water intake is 80%–90%, as per standard practice in sanitation facility design, whereas those with water intake are mostly from the literature.

Table 1: Water Use and Wastewater Estimate in a Livestock Farm

Process		Unit	Norm	No. of Heads	Volume (m^3/day)	Conversion (%)	Volume (m^3/day)
			Water Use Estimate			Wastewater Estimate	
Animal drinking and watering	Cattle	L/d/h	50	1,000	50	40	20.0
	Pigs	L/d/h	35	2,000	70	40	28.0
Washing, pen cleaning	Cattle	L/d/h	50	1,000	50	85	42.5
	Pigs	L/d/h	20	2,000	40	85	34.0
Urine, wet manure	Cattle	L/d/h	20	1,000	20	90	18.0
	Pig	L/d/h	2	2,000	4	90	3.6
Domestic water (mainly from staff)	Overnight	L/d/h	100	30	3	90	2.7
	Day only	L/d/h	60	30	1.8	90	1.6
Truck cleaning, etc.		L/truck	600	5	3	85	2.6
Total					242		153

L/d/h = liters per day per head, m^3 = cubic meter.
Source: Authors.

By logic, water use in a project is the sum of water used by the production process and water used by the humans involved. To get the former, multiply the total number of the units processed per day by the water usage per unit. For water use by humans, multiply the total number of persons by the water use norms for humans. The norms should come from the region or country concerned. If unavailable, data from literature obtained from other places that are as similar as possible can be used as proxy.

In pollution control projects such as wastewater treatment plants, wastes incineration and landfill, the central task of the feasibility study is to justify that the technology proposed can effectively reduce the pollution to the level that can meet the discharge standard (i.e., technical feasibility). Therefore, feasibility study experts have three key tasks:

(i)   collect and present data of wastewater volume and its concentration before treatment,
(ii)   estimate pollution reduction by the efficiency of the proposed technology to predict the concentration after treatment, and
(iii)   compare the predicted results with the numeric limits in applicable discharge standards.

In these types of projects, the EIA basically serves as a second opinion to verify if the feasibility study analysis is reasonable or not, and if needed, fill the gaps and propose alternatives. More details on discharge standard versus ambient standard can be found in a parallel study by the Asian Development Bank (ADB).[1]

For most projects on production or building infrastructure, pollution control is not central in the feasibility study. As a result, their feasibility study experts usually do not have expertise on environmental issues, as in the example in Box 2. The three aforementioned tasks in a feasibility study therefore fall squarely on the EIA, including data collection from second-hand sources and first-hand sample tests. In Box 2, the EIA preparers could not find a suitable, existing slaughterhouse within practical reach. Thus, they relied on second-hand sources, whose results are summarized in Table 2.

The results in Table 2 are quite revealing. First, they show the constraints facing EIA in a project setting when the EIA preparers do not have the time or resources to obtain data independently and have to rely on data from the technology providers. Second, EIA preparers usually have background in different branches of environmental science, not necessarily specialization in pollution issues. Even for those who do, their expertise may not be up to the level of enabling them to judge the efficacy of treatment technology in the feasibility study, given the sheer number of technologies and their variations.

Despite the shortcomings, the example in Box 2 and Table 2 clearly demonstrates how the standard EIA process can verify the pollution control claimed by the proposed technology and thus inform the decision by conducting the three aforementioned key tasks in the feasibility study.

[1]   ADB. 2023. Pollution Control Technologies for Small-Scale Operations. https://www.adb.org/publications/pollution-control-technologies-small-scale-operations.

Table 2: Estimate of Slaughterhouse Wastewater Treatment

Parameter	Wastewater (mg/L)[a]	DEWATS Removal[b] (%)	Results (mg/L)	Removal by CW (%)[b]	Total[c] Removal (%)	Final (mg/L)	Standard (mg/L)
BOD[a]	1,414	97	43	58	99	18	30
COD[a]	2,795	96	112	51	98	55	120
TSS[d]	886	87	115	69	96	36	50

BOD = biochemical oxygen demand, COD = chemical oxygen demand, CW = constructed wetland, DEWATS = decentralized wastewater treatment technology system, TSS = total suspended solids.

[a] COD and BOD values are from the DEWATS provider.

[b] These are DEWATS removal rates from its provider, and CW is from an internet search by the EIA preparers.

[c] Total removal rate = 1 – (1 – DEWATS removal %) x (1 – CW removal %). All figures are rounded off.

[d] Source for TSS data: single literature that the EIA preparers found due to time constraints.

The EIA preparer cannot have all-round knowledge or skills, given that demand on the profession is already overly multidisciplinary and still increasing. Many countries thus require qualification of individuals and firms by type of projects. Qualification on pollution is needed to do an EIA for animal husbandry, wastewater treatment, thermal power plant, etc. Likewise, a different qualification is required to do an EIA for flood control, dam, or highway projects that impact the ecosystem and biodiversity more prominently.

Even with the qualification, the research and estimation in EIA requires professional judgment that can only come from years of experiences and knowledge in the sector or industry. In addition, time constraints at the project level can hardly afford such researching, cross-checking, and verification of data suitability. Thus, support is needed to facilitate the task of the feasibility study and EIA preparers.

Objectives and Coverage of the Study

To fill the gaps described above, the present study was carried out and funded by a technical assistance (TA) on strengthening safeguards management in Southeast Asia (ADB 2018). Undertaken mainly by literature review and supplemented by project information, the study aims to:

(i)   Search and compile industrial norms, focusing on unit water consumption, unit discharge of pollutants (both solid and liquid), and typical discharge concentration range for main types of agriculture subsectors common in projects.

(ii)   Analyze and evaluate their reasonability and suitability for pollution quantification and prediction as needed in EIA work and in planning and selection of technologies for pollution control and treatment.

The geographic coverage is intended for developing members countries (DMCs) of ADB but much of the literature are applicable to other countries.

For the agriculture subsectors covered, air emissions are predominately fugitive emissions, i.e., odor from animal pens, manure pits, and wastewater treatment especially sludge. Although relatively harmless, it may affect quality of life of the general public and their acceptance of projects and facilities.

As for any nonpoint source pollution, quantifying odor and other fugitive emissions is difficult and rare in the literature. In addition, some of its composited gases are difficult to test especially in developing countries where the needed devices and skill are often unavailable. The emphasis should be more on their control than quantification, the main methods for which can be found in the parallel ADB study (footnote 1).

Another major air pollution stream is flue gas emission; however, this is not unique for agriculture subsectors but common to all industries that use boilers or incinerators. Given the wealth of information on flue gas emissions, e.g., the IFC's EHS general and sector guidelines on thermal power generation, there is no need for their repetition in this study.

Given the limited time and information in literature available, and more crucially, the diversity and complexity in some agriculture subsectors, this study provides only references for quantitative estimate of point-source pollution, initial screening of treatment methods, and their validity check in the EIA and feasibility study work. They cannot substitute for investigation and testing at project level.

II. Pollution Characteristics of Animal Husbandry

Water Use and Wastewater Generation in Animal Husbandry

Animal or livestock husbandry projects typically include animal farm or feedlots, breeding centers, and quarantine stations. They all need to hold animals for a certain period, depending on their different functions. Therefore, their overall water norm and resultant wastewater per unit will not vary much. This facilitates data search since some subtypes like breeding have less publicly available data compared to that of animal farms. Although most references are from an industrial scale, they are presented on a per unit basis, and hence can provide good estimates even for smaller scale operations.

Water is used in drinking, sprinkling and cooling, and washing or cleaning of pens and facilities (Table 3). For animal holding such as inspection and quarantine, cleaning must be undertaken for every new batch. In semen collection and breeding, the animal may require washing prior to each extraction. In addition to these water uses in the production process, there is general water use for the staff, sanitation, space cleaning, etc.

Animal Drinking

Water consumed in animal drinking based on a study by the Food and Agriculture Organization of the United Nations (FAO) on beef cattle in a hot environment is 30–60 liters per day per head (L/d/h). Wagners (2021) cites 53.4 L/d for cattle. For swine, Komlatsky et al. (2022) cites that water intake for boars (200 kg per head on average) is 25 L/day. In a study of drinking behavior of pigs in experimental pens, Andersen et al. (2014) estimated that 30% of drinking water was wasted and drained. Based on these, Table 3 shows that 30% of drinking water of 50 L/d/h for cattle and 25 L/d/h for pig become wastewater.

Sprinkling and Cooling

Sprinkler systems are used to cool down livestock in the holding pen and are operated only during dry or hot periods. VanDever (2017) states that the design for cattle pen sprinklers is about 20 gallons/day/head (equal to 76 L/d/h) for small capacity holding pens (40 cattle). For pigs, Fox et al. (2014) estimate a value of 7 L/d/h of sprinkling water.

Table 3: Wastewater from Livestock Holding (L/day/head)

Water Use	Water Use Norm		Conversion	Unit Wastewater Estimate	
	Cattle	Pig	Rate (%)	Cattle	Pig
Animal drinking	50	25	30	15	8
Sprinkling, cooling	76	7	70	50	5
Washing, pen cleaning	45–76	16–28	90	41–68	14–24
Urine flow			100	25	4
Total (approximate)	170–200	48–60		123–150	30–40

Sources: References cited in the "Animal Drinking" and "Sprinkling and Cooling" sections.

Water loss due to evaporation especially in tropical regions with high temperatures can vary considerably. A study by Montoya (1992) as cited in DairexNet (2019) found that 23% of water evaporates in a sprinkling operation. Calculations from poultry pens in the United States (US), however, showed that water loss due to evaporation may go up to around 50%, depending on location and climate. In this study, 30% evaporation loss is assumed, which means only 70% of water used in sprinkling can become wastewater.

Washing and Pen Cooling

For water use in pen cleaning and manure washing, the estimate of Dairexnet (2019) is about 20 gal/day/head or 76 L/d/h for cattle. Data from India is 45–70 L/d/h for cattle, 25–28 L/d/h for pig, and 36L/d/h for horse (Parihar et al. 2019). For swine, Misra et al. (2020) cited 32 liters per square meter (L/m^2) for pig washing and pens cleaning, or about 16 L/head by power washing method. Based on these, a range of 45–76 L/d/h for cattle and 16–28 L/d/h for pig are recorded for pen cleaning (Table 3). The wide range can be attributed to the cleaning methodology, hygiene standards, area per animal, etc.

Manure in pens is washed off using water, which generates a large amount of wastewater. Given that pen washing constitutes the biggest share in husbandry wastewater, dry scrubbing was developed, which greatly reduces water use and thus wastewater. Due to the much higher investment needed in scrubber systems, a hybrid method was also adopted. As reported by the Ministry of Ecology and Environment (MEE) in the PRC, wastewater from these three manure cleaning methods is, in ascending order, 10–15 L/d/h for dry scrubbing, 25–30 L/d/h for hybrid method, and 35–40 L/d/h for pen washing as reported in MEE (2009).

Not all water used will become wastewater, and not all wastewater can be intercepted into a wastewater treatment system. In sanitation facility design for processes without water intake, wastewater volume is usually calculated as 80% of water consumption. For industry wastewater generation, a higher conversion rate (e.g., 90%) is usually adopted for processes that do not integrate or absorb water.

Urine Estimation

Wastewater from animal holding also inevitably contains urine. Part of urine is absorbed in the sand bed and carried to the manure bed, yet this amount is small and can be negligible. Urine that flows into wastewater is about 28.5 L/d/h for dairy caw according to IFC's EHS guidelines on livestock production. The FAO report cited only 10 L/day pure urine for 550 kg beef cattle, although part of the manure goes with the urine; hence, the urine flow may end up at around 25 L/d/h for cattle.

For swine urine, the data from IFC GHS Guidelines for Mammalian Livestock Production (IFC 2007d) is only 1.4 L/d/h, compared to the findings of the America Society of Agricultural Engineers of 3 L/d/h and the Philippine Agriculture Engineering Standard (PAES 2001) of 4 L/d/h. Data cited by Parihar et al. (2019) indicate about 0.25–3 L/d/h of urine for pigs with average weight of 50 kg, and 8.5–23 L/d/h for cattle with average weight of 500 kg. To be more conservative considering the general situation in DMCs, 4 L/d for pigs and 25 L/d for cattle of urine are assumed.

Table 3 shows that the water norm for cattle holding is about 200 L/d/h in total, with more than a third in washing and cleaning. For pigs, the total water use is dominated by drinking and cleaning in roughly equal measure. Water use for sprinkling cattle is high even if it is only used during hot days, as hot weather prevails almost all year-round in tropical regions. General water use, such as in vehicle and floor cleaning, is discussed in chapter 3 on General wastewater estimation.

Characteristics of Animal Husbandry Wastewater

The characteristics of wastewater from cattle and pig husbandry vary among sources (Tables 4 and 5). The major reasons for these huge variations include (i) type of farm, e.g., feedlot or intensive vs. pasture, dairy caw vs. beef cattle; (ii) cleaning method, e.g., conventional washing by water vs. mechanical (so called dry-cleaning) or semidry method; (iii) scale of animal holding; (iv) feed composition; and (v) climate and temperature. While most of them are unclear about scale, the context indicates they have economies of scale, except that from Chandrasasi, Haribowo, and Wardana (2021), which only has 12 cattle. This might explain its outlying data. The sampling point also contributes to data variation. For North Dakota State University (2013), the runoff to ponds is sampled, while Daud and Anijiofojor (2017) took samples at a tank for all wastewater streams. Lastly, characteristics of wastewater also fluctuate by season and in different climates.

The parameters enumerated in Tables 4 and 5 are the significant pollutants for cattle holding and pig holding. However, intensive or superintensive systems require the use of diverse chemicals (antibiotics, algaecides, parasiticides, etc.), which also contribute to increasing the pollution, as is also true for aquaculture systems.

Table 4: Wastewater Characteristics of Cattle Holding

Parameter	Authors[a]	Daud and Anijiofojor (2017)	Othman et al. (2013)	North Dakota State University (2013)	Chandrasasi, Haribowo, and Wardana (2021)	Ministry of Ecology and Environment (2009)	PRC (2015)
BOD	440	597	1,750		64		10,000
COD	1,006	2,839	3,600		102	900–1,000	20,000
TSS	690	703	230	1,504	46		3,000
NH3-N		180		26	66	20–60	1,000
TN			650			40–80	
TP			380	75	1,219	5–20	100
Coliform		3.00E + 09					10,000

BOD = biochemical oxygen demand, COD = chemical oxygen demand, NH3-N = ammonia nitrogen, PRC = People's Republic of China, TN = total nitrogen, TP = total phosphorus, TSS = total suspended solids.

[a] Based on recent livestock projects the authors worked on in Cambodia, etc.

Notes: All units in mg/L except for coliform, which is in CFU/100ml. All blank cells mean data unavailable.

Table 5: Wastewater Characteristics of Pig Holding

Parameter	Pongthornpruek (2017)	Vanotti et al. (2014)	Nagarajan et al. (2019)	Ministry of Ecology and Environment (2009)	Giang et al. (2021)
pH	7.33–7.7			6.3–7.5	7.86
BOD	100–512		2,000–30,000	3,000–9,000	240.9
COD	270–957	9,794		15,600–46,800	505.3
TSS		6,845		2,500–4,000	162.7
NH3-N		620	110–1,650	127–1,780	
TN	17–75	1,219	200–2,055	140–1,970	126.7
TP	130–322	439	100–620		28.9

BOD = biochemical oxygen demand, COD = chemical oxygen demand, MPN = most probable number, NH3-N = ammonia nitrogen, PRC = People's Republic of China, TN = total nitrogen, TP = total phosphorus, TSS = total suspended solids.

Notes: All units in mg/L except for pH CFU/100ml. All blank cells mean data are unavailable.

Solid Waste Characteristics in Animal Husbandry

Solid wastes in animal holding are dominated by organic waste, such as manure, residues of feed, and bedding materials. Unit waste per head of major livestock species common in Asia are summarized in Table 6. Though mostly from feedlots of different livestock, these unit waste generation data can provide proxies for estimating waste from small-scale operations, at least in terms of magnitude. Depending on the pen cleaning practice, more conventional wet cleaning produces more liquid waste (a mixture of urine, manure, and wash water) than does dry or semidry scrubbing.

The Ministry of Ecology and Environment (MEE) of the PRC compiled and issued in 2020 a series of manuals on pollution generation coefficient and cross-checking methods for all major sectors and industries. One manual is for slaughtering and meat processing, which also covers animal husbandry pollution coefficient, the term used in the PRC for unit discharge of pollution. Originally developed years ago as consistent methods for census of pollution sources nationwide, they are now also used to help pollution estimation in EIA and design of pollution control and waste management facilities.

Table 6: Unit Waste Generation from Livestock Holding (kg/day/head)

Waste Sources	American Society of Agricultural Engineers (2005)	Ministry of Ecology and Environment (2020)	International Financial Corporation (2007d)	Parihar et al. (2019)	Yeo et al. (2019)
Bull, dairy caw	28	20–30	24	24	26–46
Pig manure	5	2	1.4	4	3
Sheep, goat		2.6		1.3	0.8
Broiler		0.1		0.03	0.13
Cattle bedding	2.3	2–3			

The data compiled in Table 6 largely converge on unit waste generation from cattle holding but vary for pigs (1.4–5 kg/d/h). For sheep and goat in the PRC, waste generation is double the average in Parihar et al. (2019), which provides a broad range of 1–25 kg/d/h. Data on broiler is about one-third that of the PRC , which is similar to the waste generation of the US as compiled by Yeo, et al. (2004). Data from IFC's EHS guidelines on livestock production (IFC 2007d) is from Denmark in northern Europe, which is perhaps less representative than those from the PRC and the US, as both encompass more diverse climatic zones from subtropical to temperate with more types of livestock. Feasibility study and EIA preparers need to make a judgment on which reference data better suits their tasks at hand.

III. Pollution Characteristics of Aquaculture

Water Use and Wastewater Generation in Aquaculture

One of the two major environmental issues of aquaculture is effluent pollution (the other being threat to natural habitat). This is more of a concern in tropical and subtropical areas, which have habitats comprising sensitive coral reefs and associated communities. For aquaculture in lakes, reservoirs, or seas (also called marine culture), it is difficult, if at all possible, to separate water polluted by excreta, feed, and other added materials. For effluents that can be collected for treatment, water consumption and effluents discharged can be quantified by the volume of fish or shrimp ponds, as well as interval and percentage of water replenishment required for the health and growth of aquatic products.

Operation practices greatly affect water use and effluent volume. Flow-through systems dominate developing countries whereas developed ones mostly use recirculating aquaculture system (RAS), which reduces water consumption considerably. Water use for RAS is 1–3 cubic meters (m^3) per kilogram per year compared to 30 m^3 in flow-through system (Begnballe 2015) up to 45 m^3 (Hussan et al. 1992).

General Wastewater Estimation

Water use for general purposes, also called domestic wastewater, include those needed for staff, vehicles, and facility cleaning; and in some cases, on-site laboratories. For vehicle cleaning and laboratory water use, data is hard to find and varies especially for laboratories, which can have many types, functions, target test subjects, and parameters. Quayson and Awere (2017) estimated that water use ranges between 162 L for saloon vehicles and 532 L for tipper trucks (Table 7).

Table 7: Unit Water Use for General Purposes and Associated Wastewater

Water Use	Unit	Water Norm	Collectible Wastewater	
Truck cleaning and disinfection	L/vehicle	162–532	90%	145–477
Domestic (stay overnight)	L/d/h	100	90%	90
Domestic (daytime only)	L/d/h	45	90%	41
On-site laboratory use	L/d	250a	90%	223

L/d/h = liters per day per head.

[a] Data on on-site laboratory water use is from an aquaculture project of the authors.

Source: Authors.

Generation of domestic wastewater from personnel is around 130 L/d/person according to the design code in many countries, such as in the PRC and the New Delhi building code. However, this is usually for estimating water use and domestic wastewater in a town that includes public water use in street cleaning, landscaping, etc. To estimate domestic water consumption of a facility, 100 L/d/person for overnight stay is generally accepted, and 40–45 L/d/person for daytime-only staff is commonly used (Table 7).

Characteristics of Aquaculture Effluents

Aquaculture water pollution mainly originates from excreta and uneaten feeds, hence are organic matter reflected in biochemical oxygen demand. Phosphorus is mainly from feed and fertilizer added to the pond to promote algae growth. Aquaculture effluents also contain a trace number of contaminants of emerging concern, such as antibiotics, hormones, and other veterinary medicines widely used to treat disease and improve animal health. For hatcheries, disinfectants are also commonly added to control fungal infection of eggs and to treat external fungal and parasite infection of skin and gills in the early growth stage.

The characteristics of aquaculture effluents depend on the species and systems, including operation practice used and major factors behind the variation in concentration (Table 8). The study focused on effluent data of flow-through systems, given their prevalence in developing countries with seawater or brackish water. Table 8 also includes firsthand data from shrimp hatchery projects that the authors

Table 8: Effluent Characteristics of Aquaculture (Flow-Through System)

| Parameter | Shrimp Project in Indonesia | | | Kurniawan et al. | | Tello, Corner, and Telfer (2010) | Yeo, Binkowski, and Morris (2004) | Keshem et al. (2023) |
	Ponds	Hatchery 1	Hatchery 2	Kurniawan et al. (2021a)	Kurniawan et al. (2021b)			
pH	7–8	6.7–7.8	6.9–7		6.7–7.9			
BOD	0.5–11		70–150			1–181	1–120	
COD	14–25	108–760	245–570	6.7–7.8	66–758			
TSS	8–13	0.4–45	14–94			1–201	1–300	
NH4-N	0.5–31	0.4–4.2	<0.03	108–760	0.1–25	0–1.5	0.01–9	1.5–6
NO3-N	0.5–0.7	0.8–8.5	n.a.	0.4–45	0.2–30	0–2.5		18–153
TN				0.4–4.2	0.8–18.5	0.3–3	0.2–60	11–90
TP			<0.5	0.8–8.5	0.1–32	0–1	0.01–9	1–17
PO4	1–5	0.4–4.5				0.1–0.6		8.6–16
Coliform	60,000	n.a.	<1800					
Escherichia coli				0.4–4.5				

BOD = biochemical oxygen demand, COD = chemical oxygen demand, n.a. = not available, NH4-N = ammonium, NO3-N = nitrate nitrogen, PO4 = phosphate, TN = total nitrogen, TP = total phosphorus, TSS = total suspended solids.

Note: All values are in mg/L except pH, coliform, and *Escherichia coli,* which are in most probable number per 100 milliliters.

worked on in Indonesia. Data for the two hatcheries were sampled from different tanks of their simple treatment system, which might explain the wide range in BOD, TSS and nitrogen-related parameters.

Tello et al. (2010) in the United Kingdom reviewed salmonid rearing, which also demonstrated some similarities in effluent characteristics with those in tropical regions, i.e., data from the ADB project on shrimp aquaculture in Indonesia. In two articles, Kurniawan et al. (2021a and 2021b) in Malaysia provided data on the effluents. Yeo et al. (2004) compared aquaculture wastewater with that of rural domestic and other industries. They found that the BOD and TSS of aquaculture are far below that of other industries and close to that of municipal sewage or domestic wastewater. The levels of nitrogen and phosphorus are comparable or lower than sewage.

Table 9: Comparison of Aquaculture Effluents with Municipal Wastewater (mg/L)

Parameter	Castine et al. (2013)			FAO (1992)	Ministry of Ecology and Environment (2016)	Ahmad et al. (2022)		
	RAS	Flow	MWW	MWW	MWW	Fish	Shrimp	Crab
BOD				100–200	80			
COD					220			
TSS	5–390	5–119	93–800	100–200	150			
NH4-N	6.8–26	0.41	36		22	0.6–9	1–289	2
NO3-N	10–13	0–0.23	0.01–2			0.4–30	101	4.5
TN	18	1.5–3	52	20–40	30			
TP	2.1	0.02–0.3	10	6–10	4	0.1–32		0.1

BOD = biochemical oxygen demand, COD = chemical oxygen demand, MWW = municipal wastewater, NH4-N = ammonium, NO3-N = nitrate nitrogen, PO4 = phosphate, RAS = recirculating aquaculture system, TN = total nitrogen, TP = total phosphorus, TSS = total suspended solids.

Similar findings in Australia (Castine et al. 2013) led to the recommendation of borrowing technology for pretreatment and posttreatment of municipal wastewater treatment plants to deal with the high volume of, but diluted, effluents from aquaculture systems typical in developing countries. Namely, grow-out ponds serve as pretreatment to settle bigger particles and remove them during draining and dredging after each harvest. Settlement basins can retain effluents; settle smaller particles (uneaten feed, excretion, etc.); and degrade dissolved nutrients, essentially an oxidation lagoon treatment (footnote 1). This simple approach is used by more than 70% of Australian aquaculture farms (Table 9). To meet the applicable standard, constructed wetlands are often used for further purification, much the same way as that for wastewater treatment plant discharge.

Solid Waste Characteristics in Aquaculture

Aquaculture solid waste mainly comes from excretion; leftover feeds; and erosion of pond floor, walls, and discharge channels. Yeo et al. (2004) drew a comparison with manure of livestock, poultry, fish, and human (Figure 1, part of which is also included in Table 6). On a per-wet fish basis (the first column), it is close to 0.002 kg excreta per day. However, data per ton of wet fish (second column) is about 20 kg/day. Depending on the fish species and size, this equates to 0.02 kg excreta per day per head for fish averaging 1 kg; and 0.01 kg excreta per day per head for fish averaging 0.5 kg each. To be more conservative, the second column for fish in Fig 1 is recommended, plus estimate of other smaller waste streams such as leftover feeds, etc., which again vary depending on species, technical process, and farm practice.

Figure 1: Daily Manure Production of Different Animals and Humans

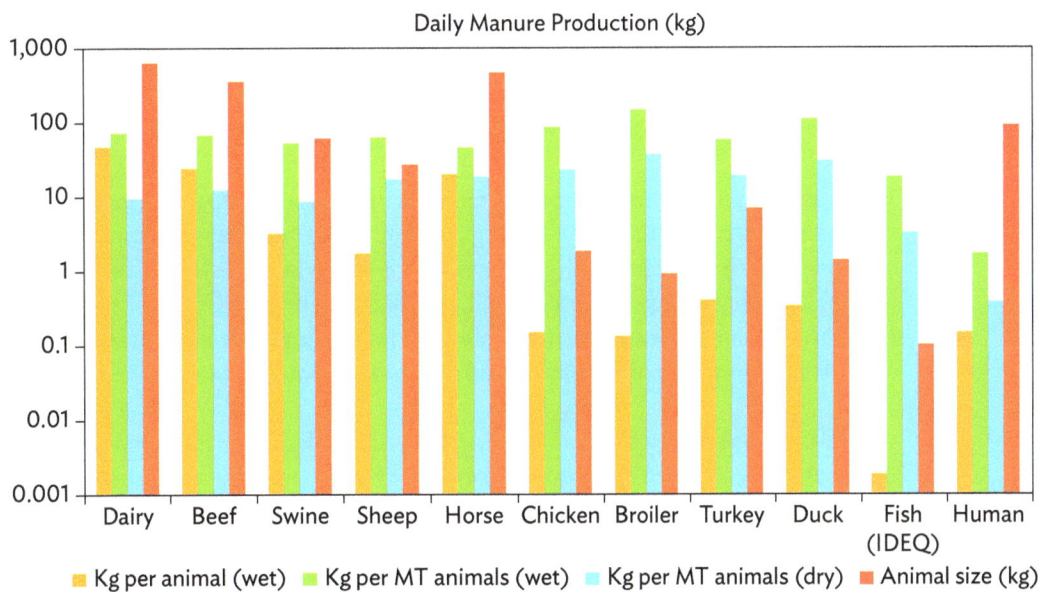

Daily Manure Production (kg)

Legend: Kg per animal (wet) | Kg per MT animals (wet) | Kg per MT animals (dry) | Animal size (kg)

X-axis categories: Dairy, Beef, Swine, Sheep, Horse, Chicken, Broiler, Turkey, Duck, Fish (IDEQ), Human

kg = kilogram, MT = metric ton.
Note: The data on fish excreta is from Idaho Department of Environmental Quality, United States.
Source: Yeo et al. (2004).

IV. Pollution Characteristics of Meat Processing

Water Use and Wastewater Generation in Meat Processing

All agricultural processing, whether animal or plants, requires cleaning of raw materials, products, and facilities including equipment, not only for production but also to meet hygiene and food safety standards. This consumes a large amount of water which is often mixed with detergents or disinfectant. Water is also used in cooling and transporting in the process

This is often mixed with detergents or disinfectant and consumes a large amount of water. Water is also used in cooling and transporting in the process.

Reuse and recycling of water is a must for economic reasons, especially if water supply is limited. Moreover, reuse and recycling can reduce the volume of wastewater and its dilution. A higher concentration of wastewater is more desirable for wastewater treatment as it lowers energy consumption per unit of pollution removed. However, reused water needs to be examined carefully if it is intended for edible food items. After all, food safety remains an overriding concern in all food processing operations.

Animal processing begins with their slaughtering. The IFC's EHS guidelines on meat processing (IFC 2007b) have much information on slaughtering steps, their health and safety risks, and measures for prevention and treatment. However, the data is only on water use per unit of carcass and rendering, not by main processing step. There is no data on wastewater generation, let alone their typical characteristics.

In slaughterhouses, water is used in rinsing carcasses, lairages, and by-products; dehairing and rind treatment of pigs; cleaning and disinfecting equipment and process areas; and livestock drinking water. In the sticking process of cattle slaughtering, some 40%–60% of the blood is recovered (Shende et al. 2022) for further processing, while the rest remains in the carcass or is washed into wastewater. Consequently, wastewater—containing organic substances, blood, hairs, and dissolved fat—is generated throughout the process, resulting in high BOD, COD, and oil and grease (O&G).

The IFC guidelines (IFC 2007b) provide a range of slaughterhouse water use for cattle ($1.6–9$ m^3/ton of carcasses) and pig ($1.6–8$ m^3/ton of carcass). These are converted to liters per head slaughtered using 1.9 kg carcass equal to 2.25 kg live animal (Sannik et al. 2015), since estimation and design are usually based on number of animals. The literature survey results are presented in Table 10.

Table 10: Water Norms for Slaughterhouse (L/head)

Source	International Finance Corporation (2007b)	Shende et al. (2022)	Bowser and Nelson (2021)	Salminen (2002)	Ministry of Ecology and Environment (2010)
Cows	405–2,250	1,114	570–1,700	1,200–1,300	1,000–1500
Pigs	200–700		171	170–700	500–700
Sheep					200–500

The variation in water use is mainly due to the size or capacity of the slaughterhouse. Shende et al. (2022) provided the correlation of water use with capacity of slaughterhouse for slaughtering 1,000 buffalo per day, a fairly large facility (Figure 2). The average water consumption is 1,127 L per animal per day, which tends to increase as the capacity decreases, indicating the economy of scale effect. According to the US Environmental Protection Agency (USEPA), implementing the Hazard Analysis Critical Control Point has led to an increase of 20%–25% in water use (USEPA 2002).

Several studies (Meat Research Corporation 1995, Enterprise Ireland 2009, Shende et al. 2022) found that 5%–20% of the water supplied is either retained in the by-products—for example, blood meal, tallow, poultry food, and waste—or lost through evaporation by heat or steam. Thus, 80%–90% of water used becomes wastewater. This corresponds well with the use of 90% by environmental engineers in estimating industry wastewater from water consumption.

Figure 2: Water Consumption vs. Slaughterhouse Capacity

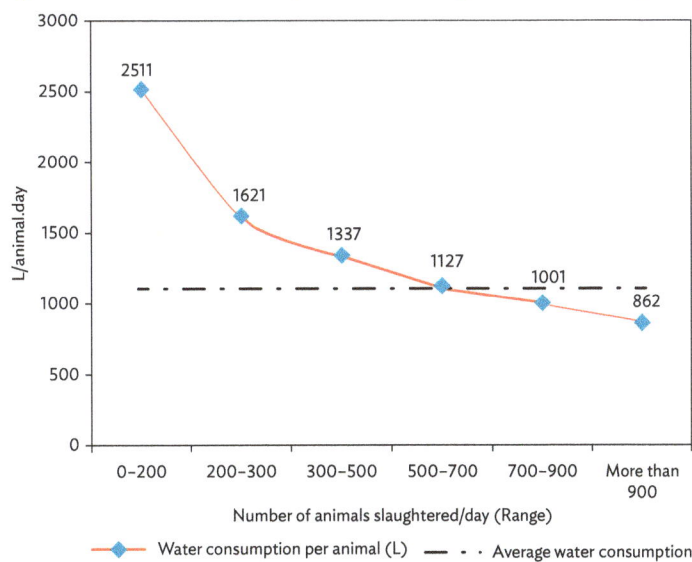

L = liter.
Source: Shende et al. (2022).

Shende et al. (2022) also provides water use during the main steps in the slaughtering process (Table 11). Paunch washing, staff hygiene, etc. are the activities that consume the most water in slaughterhouses, followed by washing and refrigeration. This water use breakdown includes the rendering process, which, however, is typically done in a separate facility. Factoring it out does not affect the overall pattern much.

Table 11: Average Water Use During Main Steps in Slaughtering (L/head)

Operation	Water Use	Share (%)
1. Lairage (6–36-hour rest before slaughtering for quality meat)	74	7
2. Sticking (i.e., beheading and mainly bleeding)	86	8
3. Fleshing (i.e., hide removal)	30	3
4. Paunch washing (removal of internal organs and content)	215	20
5. Rendering (conversion of waste tissues to usable material)	58	5
6. Carcass cutting	30	3
7. Personal hygiene, administration, canteen, etc.	210	20
8. Plant washing (of floor, equipment)	180	17
9. Refrigeration	182	17
Average water consumption per animal	**1,065**	**100**

Source: Shende et al. (2022).

For poultry slaughtering, the IFC's EHS guidelines on meat processing (IFC 2007a) presented water use data from two countries in Europe. The sources covered in the literature review by Bowser and Nelson (2021) also have water use range, similar to those of other sources (Table 12). Bingo et al. (2019) states that about 90% water used in poultry slaughterhouse water is discharged as wastewater, as water is mostly used for washing meat products and for sterilizing equipment, process, and receiving areas.

Two factors contribute to the higher water use on a per-kilogram basis for poultry compared to livestock slaughtering. One is the required continuous overflow from scalding tanks, and the other is the need for carcass immersion in ice bath chillers with a continuous overflow for removing body heat after evisceration. Livestock carcasses are chilled using mechanical refrigeration with less water use on a per-kilogram basis (USEPA 2002).

Table 12: Water Consumption in Poultry Slaughtering (L/head)

Variety	International Finance Corporation (2007a)		Bowser and Nelson (2021)		Ministry of Ecology and Environment (2010)
	Finland	Denmark	Gil and Allende (2018)	Salminen (2002)	
Broiler	17.9–18.7	16.1	13.3–38	18–18.5	10–15
Duck					20–30

Characteristics of Slaughtering Wastewater

Slaughtering wastewater typically contains a high concentration of organic material originating from processing and animal holding areas. Thus, it has high BOD and COD, nitrogen, pathogenic viruses and bacteria, and parasite eggs. Detergents and disinfectants, including acid, alkaline, neutral compounds, and pest control chemicals, might also enter the wastewater stream if applied during facility cleaning.

Table 13 shows significant variation in pollution concentration of BOD, COD, TSS, etc., the main parameters regulated by discharge standards in most countries. Oil and grease (O&G) are relevant for slaughtering wastewater but scant in the literature. The data for the PRC include mixed slaughtering of pigs and cattle (CRAES et al. 2017). Their O&G are much lower than other sources, perhaps due to the wider use of oil traps before discharge outlets where the effluent is sampled and data obtained. Other sources provided a wide range of values due to variation in sizes and production processes. Despite this, the overall characteristic remains high in organics albeit a bit lower than that for livestock husbandry.

Table 13: Wastewater Characteristics from Livestock Slaughtering (mg/L)

| Parameter | Cattle Slaughtering | | | | Swine Slaughtering | | | Mixed |
	Lvfeng Environmental Engineering Company	Ziara, Subbiah, and Dvorak (2018)	McCabe et al. (2013)	Nwuba and Orakwe (2019)	João et al. (2020)	Park, Oh, and Ellis (2012)	CRAES et al. (2017)	CRAES et al. (2017)
BOD	1,000	1,486 ± 831	163–7,020	1,049	3,018	5,732 ± 1,522	500–1,200	420–1,900
COD	1,500–2,500	4,185 ± 2,141	1,040–12,100	1,699	4,380	7,864 ± 4,294	1,500–2,000	1,200–3,900
TSS	1,000	4,973 ± 2,526		886	1,000	2,355± 1,321	400–1,000	950–1,300
O&G	50–200	269 ± 196	2,110		100		40–65	25–50
NH3-N	100–150						15–120	

For poultry, Table 14 shows great variation of pollution concentration in BOD, COD, TSS, and O&G due to similar reasons. Some of the variations also reflect different levels of effort among plants to minimize water use and reduce the cost of wastewater treatment. Uncollected blood, solubilized fat, and feces are principal sources of BOD in poultry processing wastewaters. As with livestock slaughtering and meat processing wastewaters, the efficacy of blood collection is a significant factor in determining BOD and nitrogen concentration in poultry processing wastewaters (USEPA 2002).

Table 14: Wastewater Characteristics from Poultry Slaughtering (mg/L)

Reference	Country	BOD	COD	TSS	O&G	NH3-N
Aziz et al. (2018)	Malaysia	573–1,177	777–1,825	395–783	2,362–3,616	
Rajakumar et al. (2011)	India	750–1,890	3,000–4,800	300–950	800–1,385	
United States Environmental Protection Agency (2002)	United States	1,662		760	665	
Ministry of Ecology and Environment (2010)	People's Republic of China	750–1,000	1,000–2,000	750–1,000	50–200	50–150

BOD = biochemical oxygen demand, COD = chemical oxygen demand, NH3-N = ammonia nitrogen, O&G = oil and grease, TSS = total suspended solids.

Solid Waste Characteristics in Slaughtering

The meat processing industry slaughters animals to produce primary carcass products, processed cuts, and a variety of by-products. The rendering industry processes the parts not used for human consumption into animal feed, etc. Both generate large quantities of solid waste and by-products that can generally be divided into the following categories: (i) manure and gut content, (ii) oil and fat recovered by fat-separators, and (iii) hazardous waste including special risk materials from cattle slaughtering.[2]

Most of the solid waste in slaughterhouses is generated at the reception to the evisceration step. Those that can be used as by-products are mostly from stunning, dehairing, and removal of usable offal and hide. Hazardous wastes are generated mostly during meat inspection and evisceration. For the cattle slaughtering process, experts estimated hazardous wastes at about 30% of total wastes, based on their experience. That share for pig is much smaller than cattle slaughtering largely because the latter has special risk materials.

According to EHS guidelines on meat processing, in most developed countries, the by-products from cattle may exceed 50% of the animal's live weight and 10%–20% for pigs. In developing countries, utilization of animal parts is often higher, including intestines, blood, dung, bones, and hooves, likely leading to a bit lower waste to animal ratio. Their unit waste generation from different literature are collected in Table 15.

---

[2] Defined in the EHS guidelines on meat processing, special risk materials (SRM) are tissues in cattle that contain the agent that may transmit bovine spongiform encephalopathy, transmissible spongiform encephalopathy, or scrapie disease if reprocessed into animal feed. The human disease may result from human consumption of products from animals infected with bovine spongiform encephalopathy. Although not typically used for food, processing activities may accidentally result in the mixing of SRM tissue with meat products produced for human consumption. Therefore, SRM should be carefully separated from carcasses before their processing into commercially valuable by-products, whether for human or animal consumption.

Table 15 shows some similarity among different sources of data on total unit waste per cattle slaughtered, although the data in the PRC is much lower. For pig slaughtering wastes, data from the US and the PRC are similar: both are more than double that in the IFC's EHS guidelines for meat processing, which is from the Nordic countries council.

Table 15: Unit Generation of Solid Waste from Slaughtering (kg/head)

Sources		American Society of Agricultural and Biological Engineers (2005)	International Finance Corporation (2007b)	Experts	Ministry of Ecology and Environment (2020)
Cattle	Gut contents and/or waste	20		4.8	
	Manure			22	
	Total waste	50	58	50	30
	By-products for rendering		110		
Pig	Stomach contents and/or waste	1.6			
	Manure	5			
	Total wastes	6.6	2.2		5.2
	By-products for rendering		21		
Broiler	Total wastes				0.2
Duck	Total wastes				0.3

V. Pollution Characteristics of Fish Processing

Water Use and Wastewater Generation in Fish Processing

Fish processing refers to both seawater and freshwater fishing products, ranging from fishes, shrimps and lobsters, crabs and clams, etc. It mainly includes eviscerating, cleaning, filleting, injecting carbon monoxide, storing, chilling, and packaging. Preprocessing may be carried out on board fishing vessels, such as eviscerating tuna and beheading shrimps at sea with wastes disposed directly into the sea.

Fish processing wastewater is high in organic content, reflected in high BOD and O&G, and generally lower on nitrogen content. BOD is derived mainly from the eviscerating process and general cleaning, and nitrogen originates predominantly from blood in the wastewater stream. Therefore, as with meat processing, removal of blood, scale, internal organs, and their content from wastewater can greatly reduce its strength and thus make its treatment easier and less costly.

Characteristics of Fish Processing Wastewater

The results of the literature survey in Table 16 show the great variation in key pollution parameters from fish processing, more than meat processing in general. For 5-day BOD (BOD$_5$), data from most sources fall within the range of 500–4,000 mg/L despite some very high outliers. The BOD–COD ratio is around 0.5, indicating good biodegradability for standard secondary treatment. In most literature surveyed, TSS, ammonia nitrogen, and phosphorous vary more than BOD and COD.

Pollution characteristics by major aquatic species commonly processed are more available in the literature (Tables 17 and 18) than by major production process and steps. Though wastewater from filleting is lightly polluted, its pretreatment steps like washing and scaling are more polluting and water-intensive, thus generating more wastewater, as exhibited in Table 18.

Both tables indicate that canning is another water-intensive process and is therefore a wastewater source. Without canning—which is the case in most small and medium-sized enterprises—water use ranges from 10 to 20 m^3/ton of raw material processed. Similarly, some cleaner production programs found water use is about 5–11 m^3 for filleting, 15 m^3 for canning, and 0.5 m^3/ton of fish for fish meal (UNEP 2000). Water use for fish filleting is about 5–10 m³/ton raw fish (IFC 2007c). Carawan, Chambers, and Zall (1979) found that 5–10 m³/ton processed is typical of large operations with semiautomation and water-saving practices.

However, interviews at fish-filleting plants in some Southeast Asian countries by the authors revealed that water consumption is roughly 1 m³/ton of fish processed, much lower than the literature. These plants are small, with processing capacity of 10–50 tons of raw fish per day. Their lower water use might be due to lower standard on hygine etc, given the large baseline water needed for cleaning to meet hygiene requirements.

In general, wastewater volume is high for rinsing, cleaning (process, equipment, and floor), and canning. It is medium volume for filleting, and low for blood water from degutting, etc. In terms of pollution load, the sequence can be the opposite (Tay et al. 2022; Table 18).

The species processed and product produced, the scale of operation, the types of processed included and technology used, as well as the level of water minimization in place are major determinants of water use and wastewater volume. Greater variation in these reflects the greater diversity of raw materials; consequently, their products dealt with by fish processing. Reducing wastewater tends to significantly reduce organic load, as the methods typically reduce product contact and better segregate high-pollution streams.

Table 16: Wastewater Characteristics of Fish Processing by Source

Parameter	Cristovao et al. (2012)	Cristovao et al. (2015)	Muthukumaran and Baskaran (2013)	Ribeiro and Naval (2017)	Ferraciolli et al. (2017)	Kurniasih et al. (2018)	Islam et al. (2004)	Anh et al. (2021)	Ching and Redzwan (2017)
pH	6.3–7.0	6.13–7.14	5.5–8	6.0–7.0	5.5–8.5			6–11	6.65
BOD	1,129	463–4,560	2,500–3,500	463–4.569	487–1,350	500	500–1,550	500–1,500	18,419
COD	1,967–21,821	1,147–8,313	1,518–2,900	1147–8313	110–1,722	1,056	400–2,000	1300–3,250	30,000
TSS	324–9,407	324–3,150	150–3,000	324–3150	60–940	57	100–800	150–1,100	5,530
TKN	98–211	21–471	112–347	21–471	10.8–102			350–1,200	
NH3-N	3.2–19.4	3.2–1,059				1.1		8,200	
TP	16.6–67	13–47	197–291	13–47	16.4 (max)			13–47	95.5
Fat, O&G	409–2,841	156–2,808	409–2,841	156–2,808		8		156–2,808	40–300
Coliform			1–1,000		1–1,000				
Wastewater (m³/ton processed)	27							5–30	

BOD = biochemical oxygen demand, COD = chemical oxygen demand, m³ = cubic meter, O&G = oil and grease, NH3-N = ammonia nitrogen, O&G = oil and grease, TKN = total Kjeldahl nitrogen, TP = total phosphorus, TSS = total suspended solids.

Note: All values are in mg/L, except for pH and coliform, which are in most probable number per 100 milliliters.

Table 17: Wastewater Characteristics of Fish Processing by Species

Parameter	Shrimp	Crab	Clams	Oysters	Scallops	Herring	Tuna	Salmon	Range	Reference
BOD	2,000	4,400	500–2,500	250–800	200–1000	1200–6000	700–750	250–2,600	200–6,000	Islam et al. (2004)
	720–1,100						500–1,500			Anh et al. (2021)
										FAO and WHO (2023)
COD	3,300	6,300	1,000–4,000	500–2,000	300–1,100	3,000–10,000	1,600	300–5,500	300–10,000	Islam et al. (2004)
							1,300–3,250			Carawan, Chambers, and Zall (1979)
	1,200–2,300									Alexandre et al. (2011)
TSS	900	620	600–6,000	200–2,000	1,000–4,000	600–5,000	500	120–1,400	120–6,000	Islam et al. (2004)
	122–872									Alexandre et al. (2011)
TKN									347–1,200	Picos-Benitez et al. (2019)
	45–77									Anh et al. (2021)
	150 (canning)									Islam et al. (2004)
NH3-N	650 (canning)									Islam et al. (2004)
TP									13–47	Cristovao et al. (2015)
	18–71									Anh et al. (2021)
Fat, O&G	700	220	20–50	30	15–25	600–800	250	20–550	15–800	Islam et al. (2004)
Water use (m³/ton) processed	15–84						10–20			FAO and WHO (2023)
	17					7–8				Holland (2018)

BOD = biochemical oxygen demand, COD = chemical oxygen demand, FAO = Food and Agriculture Organization of the United Nations, NH3-N = ammonia nitrogen, O&G = oil and grease, TKN = total Kjeldahl nitrogen, TSS = total suspended solids, WHO = World Health Organization.

Note: All values are in mg/L except indicated otherwise.

Table 18: Wastewater Characteristics by Main Fish Production Process and Step

Parameter	Rinsing	Scaling	Fileting	Cleaning	Freezing	Canning	Fishmeal	Cooking	Reference
BOD	590	560	860	100–1,546	240	1081 (shrimp)	100–24,000		Islam et al. (2004)
		35 (white); 50 (oily)			130 (shrimp)	52 (fish); 15 (tuna); 120 (shrimp)	30,000		Venugopal and Sasidharan (2021)
				788.33				971	Kurniasih et al. (2018)
						463–1569 (fish)			Ferracioli et al. (2017)
COD				2,518		2,296 (shrimp)	150–42,000	1,022	Kurniasih et al. (2018)
						1,795–2,053			Islam et al. (2004)
				1,967					Ching and Redzwan (2017)
			50 (white); 85 (oily)			116 (fish)	50,000		Venugopal and Sasidharan (2021)
								66,222	Uttamangkabovorn et al. (2005)
						1147–8,313			Cristovao et al. (2015)
TSS	380	2,590	860	90		650–9,407			Islam et al. (2004)
						324–3,150 (fish)			Ferracioli et al. (2017)
						51–91 (tuna)			Jemli et al. (2015)
						5 (fish); 4–17 (tuna); 54 (shrimp)	30,000		Venugopal and Sasidharan (2021)
TKN				211		98–196 (shrimp)			Islam et al. (2004)
						21–471			FAO and WHO (2023)
NH_3-N						650			Islam et al. (2004)
TP or PO_4						13–47			Cristovao et al. (2015)
Fat, O&G	100	300	710	30–409		258 (shrimp)	20–5,000	1,727	Islam et al. (2004)
									Uttamangkabovorn et al. (2005)
Water use (m³/ton) raw fish processed	5.5	3.2	4.6	0.16	7.0 (shrimp)	5–15 (tuna)	0.5		Islam et al. (2004)
	10–12 (handling)	10–15 (white fish); 0.2–0.9 (oily fish)	5–11 (white fish); 5–8 (oily fish); 1–2 (ungutted); 5–8 (oily fish)							Venugopal and Sasidharan (2021)
						13 (tuna or pet food)			Ribeiro and Naval (2017)
									Uttamangkabovorn et al. (2005)
					10 – 30	15–30	12		Gómez-Sanabria et al. (2020)

TP = total phosphorus, TSS = total suspended solids, WHO = World Health Organization.
Note: All values are in mg/L except indicated otherwise.

Solid Waste Characteristics in Fish Processing

As highly perishable commodity, more than 70% of the fish caught undergoes some processing, resulting in large amounts of offcuts and wastes (Uttamangkabovorn et al. 2005). The major types are blood, offal products, viscera, fins, fish heads, shells, skins, etc. They are only partly used for the production of fish meal, fertilizers, and fish oil; or utilized as direct feeds in aquaculture. The rest is thrown away or hauled by collectors for a fee, as discovered by the authors during project field work. Insufficient amount of offcuts from small plants often cannot offset the investment and operation costs from making wastes into by-products.

The amount of waste varies depending on the level of processing (e.g., gutting, scaling, filleting) and species. These operations generate discards, which mainly include muscle trimmings (15%–20%), skin and fins (1%–3%), bones (9%–15%), heads (9%–12%), viscera (12%–18%), and scales (5%) (Coppola et al. 2021). According to Zachritz and Malone (1991), viscera is 15%–25% of the processed whereas total waste is 30%–40% of raw fish processed. In a review by Muthukumaran and Baskaran (2013), waste from filleting is 65% and from fish steak is 30% of raw weight.

Waste generation also depends on species processed, because each specie has a specific composition, size, shape, and intrinsic chemistry. For example, from the authors' field investigation in several fish processors in Indonesia, tuna filleting usually generates waste (head, bones, fins, etc.) of about 50%–60% of wet weight. Without heads, total waste is about one-third to 40%. This means tuna head accounts for about 20% of raw fish weight. Muthukumaran and Baskaran (2013) provided other waste per weight of raw material by species: 50% for shrimp, 20%–30% for whole squids or cuttlefish, and 50% for tubes of squid or rings of cuttlefish.

Size of raw materials can affect waste amount not only through processing methods but also fishing practice, which in turn has a bearing on waste amount and types in processing plants. For example, big fish like tuna is usually degutted on board before landing in the processing plant, whereas smaller fish for home consumption are just frozen on board without removing viscera, etc.

VI. Pollution Characteristics of Fruit and Vegetable Processing

Water Use and Wastewater Generation in Fruit and Vegetable Processing

Fruit and vegetable processing include trimming, sizing, peeling, blanching, fermenting, pureeing, cooking and canning, pickling, dehydrating, powder making, etc. Depending on the raw produce and products, production can involve all or just a few of the above methods, but all require basic steps of washing and cleaning to meet hygiene and food safety requirements. The different processing types illustrate the diversity of plant-based processing, in addition to variety of species and cultivar processed, end products, and operation practice such as the extent of water recycling.

All these variables lead to huge differences in water use and wastewater characteristics in terms of volume, main pollutants, and their concentration range. Therefore, the information in this chapter should be treated and used as a starting point and cross-reference to the EIA and initial screening of effluent treatment methods. More accurate and representative data of wastewater from the type of food processing in question still come from sampling and testing. Data in this report can help to validate the reasonability of test results, as plant-based processing wastewater does exhibit some common features different from animal-based wastewater, i.e. carbohydrates (e.g., sugars, starches); pectin; lignin or tannin, etc., and often low in nitrogen.

For vegetable processing, Bosak et al. (2016) studied potato farms in Canada where the on-farm processing mainly includes washing, sorting, and transporting, very much resembling those in small and medium-sized enterprises (SMEs) in developing countries. Its wastewater strength is lower than industrial-scale processing, which involves peeling, cutting, and precooking (Table 19).

Lehto et al. (2014) studied root vegetables (carrot) and lettuce (for salad) processing in Finland. They found that water consumption and wastewater strength are much higher in peeling than simple processing (wash and pack), and understandably higher for root vegetables. Water use falls into the following categories: peeling (42%), washing (30%), cleaning premises and equipment (13%), and others (14%). Peeling, etc. contribute 90% of BOD. If this can be treated separately, the rest of wastewater can be reused and treated more easily thus achieving higher efficiency overall.

Table 19: Wastewater Characteristics of Plant-Based Processing

Parameter	Rodrigues et al. (2022)	Amor et al. (2012)	Puchlik (2016)	Valta et al.	El-Kamah et al.	Bosak et al.
	Portugal	Portugal	Poland	Greece	Egypt	Canada
	Fruits (not specified) juice or syrup	Pear, apple juice	Fruit juice, vegetable pureeing	Peach, tomato syrup	Apple, orange, and cherry juice	Potato on-farm washing and cleaning
pH	11.8+/-0.4		4.3–7.9	6.5–8	5.4–8	
BOD	1520+/-30	13,900	860–3,200	1,130	3,134–+/-1,546	1,200-1,700
COD	6,400+/-300	21,040	919–3,700	2,250	5,157–+/-2,897	
TSS		3,130	249–420	70	323–+/-349	3,200-6,700
TN	14+/-3		40–60		58.2–+/-59	311
NH3-N						57–244
TP		512.4	9.4–16		10.2–+/-5.3	22–55

Note: All values are in mg/L except pH.

Based on edible parts, fruit processing can be generally grouped into utilizing the flesh (citrus, mango, apple, etc.) and the seeds (various nuts, coffee, cacao, etc). For the former, Puchlik (2016) studied the industry in Poland, which is dominated by simple processing (wash, sort, and pack: 88%) for direct sale and the rest is mainly juice. This corresponds well with the authors' findings in projects that support SMEs in agriprocessing. Most of their outputs are for direct consumption locally or for export, after simple cleaning, sorting, and packaging. Only the off-specs are further processed into frozen or dried cuts, puree, jam, juice or concentrate, etc. Wastewater characteristics by major types of processing methods are gathered to the extent possible from various sources (Table 20).

For processing that utilizes the seeds, coffee and cacao are prominent for tropical regions and thus often involved in ADB projects. Coffee processing has two stages. The first is cherry-to-green coffee and must be near the farm soon after the harvest to minimize degradation in quality. This stage generates practically all wastewater and most solid waste. The second stage includes dehulling and roasting with little pollution and done in cities.

According to the processor interviewed by the author, cherries are first floated in water to remove empty ones and dirt, using about 0.5 m³/ton of cherry with recycling. The cleaned cherries are de-pulped by simple machine. The resulting beans are fermented (for 5–7 days), which is needed to enhance quality and to ease the removal of parchment, mucilage, etc. with water. Wastewater is dripped during fermentation, often untreated especially when fermentation is done in the household. After fermentation, beans are washed, generating the strongest wastewater. The so-called green coffee is ready for dehulling and roasting after drying by sun and wind for another week.

The type and amount of wastewater depend on the method of processing coffee cherries, i.e., dry, wet, and semi-wet. Semi-washing and full wash use between 3–4 m³ and up to 14–17 m³ water, respectively, per ton of fresh coffee cherries (WorldAtlas). Water used in depulping of cherries is the largest amount in coffee processing, accounting for over half of the water use (same source). In the coffee processing studied by the authors in a project, the total wastewater from cherry to green coffee is about 1.5 m³/ton cherry according to the processor, much lower than the literature, perhaps due to its cottage industry nature and lower hygiene standards. However, such wastewater is highly concentrated with mucilage, acidic and viscous, posing difficulty to its treatment.

When reusing sewage for irrigation of vegetable, the bacteria indicators should meet applicable standards to avoid food contamination. The World Health Organization guidelines (2005) set the limit of fecal coliform <1,000 CFU/100 ml for wastewater that can be reused in irrigating vegetables. Typically, tree and berry fruit wastewater have a very high BOD-to-nitrogen ratio. It thus may be advantageous to add a nitrogen source to facilitate its treatment through biodegradation.

Table 20: Wastewater by Major Processing Types of Fruit and Vegetable

Produce	Vegetable			Utilize Flesh of Fruit			Utilize Seed	
Process	Washing	Peeling	Canning	Pureeing and juicing		Canning	Coffee	Coffee
Source	Lehto et al. (2014)	Lehto et al. (2014)	Ghangrekar. com	Zema et al. (2019) – citrus	Mohsen et al. (2012) – mongo	Ghangrekar. com – citrus, apple	Viet Nam	Cardenas et al. (2009)
Parameter								
pH				3.3–5.5	5.2		5–5.6	4.2
BOD	1,100	6,500	400–1,800		720	2,100–3,000/ 1,600–3,400	1,100–3,200	
COD	5,200	9,700		5,000–27,000	815		3,100–4,200	3,818
TSS	10,000	1,300	550–900	3,500	60–80 (122NTU)	1,700–3,400/ 300 (apple)	700–870	
TN	77	68			5.2			
TKN				60–300	720			
TP	16	45			815		5–6.5	
Coliform	>1M	>1,000			60–80 (122NTU)			
Escherichia coli	~10,000	~500			5.2			
Water (m³/t of raw)	6.5[a] (IFC 2007d)[a]			1–17	720	3–4 for durian, jackfruit, etc. (authors)		30–40 (Ijanu et al. 2020)

Note: All in mg/L except indicated otherwise. Total coliform and *Escherichia coli* are both in most probable number per 100 milliliters.

[a] It is unclear if this water norm of 6.5 m³/t is by raw produce or product.

Solid Waste Characteristics in Fruit and Vegetable Processing

Solid wastes mainly are peels; shells; stones; leftovers from extraction of oil, sugar, sap, etc. Most of the solids are organic and can be utilized to make by-products. From the literature and authors' experience in projects, half or more of fruit by weight become wastes in processing, depending mainly on species, variety, tree age, and even elevation (Table 21). Those with thick husk and those utilizing seed can have a higher percentage of wastes by weight of raw fruit processed.

Cacao processing is similar to that of coffee. Raw pods are removed manually, exposing beans in white pulp (about 30% by weight of raw pod). They are spread on banana leaves for natural fermentation (5–7 days) for similar reasons as coffee fermentation. During the period, pulp drips are collected, which quickly turn to vinegar under tropical temperature and usually for local consumption. After fermentation, beans are washed, the main step generating wastewater. Beans are then dried in a similar way as coffee beans and reduces to 30% by weight for white beans, i.e., 10% of pod weight. The dried beans are dehulled and roasted into final product, which is about 65%–70% of dry bean, i.e., 6%–7% by weight of raw pod.

Table 21 : Characteristics of Wastes from Major Tropical Fruit Processing

	Durian, Jackfruit	Mango	Citrus	Pineapple	Coffee	Cacao
Major process	Fresh cut by peeling and cleaning	Pureeing to make juice, etc.	Juicing and producing marmalade	Canning, juicing, etc.	Washing, fermenting roasting	Husking, fermenting, winnowing
Sources	Field work of authors	Owino and Ambuko (2021)	Zema et al. (2019); Suri et al. (2022)	Sarangi et al. (2022)	Ijanu et al. (2020)	Vásquez et al. (2019)
Waste (% weight of fresh fruit)	Peel: 60%–70%; seed: 10%–15%	30%–50% by cultivar	Peel: 50%–60%; seed: 15%	60% (peel, core, etc.)	60% of cherry (90+%, from authors)	80% of pod (90+%, from authors)

As with all agriprocessing, the more intensive the processing, the more value added; however, the more waste is generated as well, making it imperative to turn wastes (liquid and solid) into by-products. However, making by-products often entails economies of scale to afford the more complicated processes, more costly equipment, as well as associated operation skills, which are often beyond the reach of individual SMEs. Solid wastes (husks, peel, etc.) are simply piled to turn into compost or fertilizer; and seeds are used to grow seedlings and collect pulp, etc. as they are unsuitable for human consumption or animal feed.

VII. Discussion and Conclusions

Throughout the research for this study, it became clear why the prestigious environmental, health, and safety (EHS) guidelines of the IFC did not include water norms and quantitative pollution characteristics except for a few. These agriculture subsectors are too diverse in raw materials (i.e., species and cultivar), end products, produciton methods and technology, as well as operation practices, all leading to vast variation in unit water use, unit discharge, and concentration range. Only livestock raising and processing handle fewer species, use similar methods to raise and slaughter, and thus result in less varying water and pollution norms, as evident in this report.

Such variations not only have rendered compilation of water norms and unit discharge difficult, but also made it hard to predict pollution and wastes based on the reference values compiled here. Therefore, they cannot substitute for case-to-case sample testing of water pollution, air emission, and solid wastes for a project in question.

For a greenfield project, one needs to investigate existing facilities in the same industry or subsector, preferably of similar scale and management practice in as similar circumstances as possible. By extrapolation and analogy, one can roughly predict the new project's pollution. If such proxy facilities have pollution treatment and use technology pondered by the new project, they can also shed light on treatment performance, experience, and lessons, and help to decide their suitability for the new project.

However, the reality does not always afford this approach, especially in smaller developing countries or regions where it is hard to find existing industries or facilities that can meet the above criteria as proxies for feasibility study and environmental impact assessment (EIA) work. Even if this is not the case, it is always useful to cross-check findings from investigation with industry norms, unit discharge, and typical pollution characteristics, which this study contributes.

EIA preparers constantly face a lack of information in the feasibility study or elsewhere. In fact, nearly all professions have to get the most out of what little is on hand. On the other hand, poor or scant feasibility studies offer more room for EIA to help optimize the project design and technological choices, e.g., by proposing methods and technology to treat pollution if it is virtually absent in the feasibility study. Moreover, in most countries (explicitly so in many DMCs), EIA is required to be conducted in parallel with the feasibility study, not after. Once the feasibility study is completed, it becomes too late to modify the project design, and choice of production process or pollution control—in short, too late to inform the investment decision.

In reality, balancing the iterative process between a feasibility study and EIA to inform each other demands professional judgment and experience. It also depends on the preparation of the feasibility study, as the EIA needs some basic features about a project to assess. The feasibility study usually begins with technical feasibility analysis, followed by financial and economic analysis. The former entails technical design, which is the basis for every assessment (including EIA) that is required before the decision.

Even if the draft feasibility study does not have technical processes yet, the standard one(s), or those in the literature for that sector or industry, be it production or pollution treatment, can be used to draw indicative flowcharts and foresee pollution. However, for more diverse and highly varying sectors like processing of fish, fruit, and vegetable, EIA preparers hardly have any standard production processes and must wait for the feasibility study to provide basic features of the project.

Despite the constraints, being proactive in identifying different technologies for pollution control or cleaner production is central for alternative comparison in EIA. Some even consider the EIA process as about finding and comparing alternatives that have less adverse consequences. When the feasibility study is largely ready, it is more difficult to adopt better alternatives that the EIA recommends.

However, what the standard EIA methods described in chapter 1 usually quantify is pollution at discharge points, either with or without treatment. From this point, pollution will undergo dispersion, degradation, absorption, etc. as it travels through environmental media like air and water. The form and concentration of pollution at the receiving end are affected by many factors that can hardly be captured by stoichiometric estimation in chapter 1. Computer-based models have been built to undertake such tasks to predict concentrations at receptor end to see if they exceed the environmental quality standard or not.

Sophisticated computer-based models require a large amount of quality data. They often resemble a black box in the sense that even the modelers cannot foresee the likely results or even trends for any given input data, let alone EIA preparers. The situation can deteriorate as the model becomes more complex and covers more factors (Box 3). What EIA professionals can do is to employ standard EIA methods and practices to provide, as best as they can, the quantitative estimate of pollution at discharge to lay down a solid basis for computerized models. In cases where only compliance with a discharge standard is regulated, as prevails in many developing countries, estimating pollution at discharge in EIA is more crucial.

Box 3: The Myth of Computer Models vs. Other Methods to Predict Impacts

Some environmental professionals like those in multilateral development banks tend to get carried away by computerized models in impact assessment and often require EIA preparers to use them. One fact needs to be highlighted first: any modeling begins with pollution level (by concentration or load) at discharge point. The analysis methods described in this report are exactly for that. Hence, they are indispensable for obtaining the needed input to models. Otherwise, "garbage in, garbage out."

Second, most computerized models are designed to predict concentration at different distances from the discharge points after dispersion and various reactions through air or water, etc. In other words, models are often used to predict the impacts at the receiving end in the ambient. The meteorological, hydrological, chemical, photochemical, or biochemical, etc. all affect dispersion and transformation of pollutants in the process. Such multifactor simulation required is beyond human mind, thus the need for computers.

Majority of EIA professionals are not modelers; hence, they have to trust models, which are chosen or recommended in some countries by EIA authorities, or follow technical guidelines. What EIA professionals can strive for is to ensure that their prediction of pollution at discharge points are as robust as possible. This can be achieved by the standards and good practices presented in this report.

Moreover, affordable pollution treatment methods especially in rural settings are usually simple and cheaper technologies that might not be able to meet discharge standards. Thus, further land treatment is needed, e.g., artificial wetland followed by land application. As a result, there is no direct discharge to the water bodies for EIA or models to simulate.

However, computers also have limits and cannot beat simple methods in some cases. An example is urban roads with noise coming from diverse, existing sources, in addition to road traffic. It has proven to be very difficult, if indeed possible, for computerized models to capture all the variables and mechanisms to arrive at a reliable prediction of future noise. In such cases, investigating the combined noise and impacts of roads with similar circumstances and traffic as analogy and proxy can often lead to better and quicker prediction at much lower cost.

Source: Author.

With decades of public and regulatory pressure such as the EIA requirements, some environmental dimensions have been mainstreamed in technical design. For example, noise prediction and mitigation has long become a must in feasibility studies for transport projects that are near or through urban areas and rural settlements, even for many middle-income developing countries. Likewise, prediction of air emissions and measures to control have been integral to feasibility studies of thermal power projects in many developing countries. Therefore, EIA in these cases largely has played a role of validating and supplementing.

As a logical result, if anything goes wrong, e.g., noise from a road built in a project causing complaints and grievances, one should not automatically look to its EIA but first and foremost, check its feasibility study, in particular its traffic forecast and the resulted noise prediction. If the feasibility studies are sound and consistent with the EIA, the next step is to check if the detailed design truly reflects the feasibility study and EIA regarding noise and its control. The final step is to check the execution of the design, i.e., if it is built accordingly. Any of these steps can fall short, the consequences of which, however, fall either on the environment or people. Thus, EIA or more broadly, "safeguard" as termed in ADB, often bear the blame.

The ADB safeguard policies aim to avoid and minimize adverse environmental and social consequences of its projects and operations. Such objective can only be realized through the project design (from feasibility study to detailed design) and execution (i.e., construction and operation). As stated above, each step on the technical side of the project cycle can derail no matter how good the impact assessment and their action plans are. Tracing back to key steps on the technical side and tightening them can better foster mainstreaming environmental–social considerations in project design and execution than mere strengthening EIA and safeguards.

Even for projects where feasibility studies do not play a primary role in pollution prediction, technical experts inevitably assume ultimate responsibility, as they are in the position and have the expertise to ensure that detailed design and execution follow the feasibility study and EIA. The fact that technical issues are normally hard for other professionals and the public to grasp is perhaps one reason for more public attention on things like EIA. This adds push for ADB to improve EIA and its implementation while being aware of the constraints faced.

References

ADB. 2018. *Technical Assistance for Strengthening Safeguards Management in Southeast Asia.* Manila. https://www.adb.org/projects/documents/reg-52059-001-tar.

ADB. 2023. *Pollution Control Technologies for Small-Scale Operations.* Manila. https://www.adb.org/publications/pollution-control-technologies-small-scale-operations.

Ahmad, A. L. et al. 2022. Environmental Impacts and Imperative Technologies towards Sustainable Treatment of Aquaculture Wastewater: A Review. *Journal of Water Process Engineering.* 46. 102553.

Alexandre, V. M. F. et al. 2011. Performance of Anaerobic Bioreactor Treating Fish-Processing Plant Wastewater Pre-hydrolyzed with a Solid Enzyme Pool. *Renewable Energy.* 36 (12). pp. 3439–3444.

American Society of Agricultural Engineers (ASAE). 2005. Manure Production and Characteristics. Michigan.

Amor, C. et al. 2012. Treatment of Concentrated Fruit Juice Wastewater by the Combination of Biological and Chemical Processes. Part A: Toxic/Hazardous Substances & Environmental Engineering. *Journal of Environmental Science and Health.* 47. pp. 1809–17.

Andersen, H. M. L., L. Dybkjær, and M.S. Herskin. 2014. Growing Pigs' Drinking Behaviour: Number of Visits, Duration, Water Intake and Diurnal Variation. *Animal.* 8 (11). pp. 1881–1888.

Anh, H. T. H. et al. 2021. Options for Improved Treatment of Saline Wastewater from Fish and Shellfish Processing. *Frontiers in Environmental Science.* 9 (689580).

Ark Viet Nam. 2023. Vietnam: Standard and Modern Coffee Processing Wastewater Treatment Methods.

Aziz, H. A. et al. 2018. Poultry Slaughterhouse Wastewater Treatment Using Submerged Fibers in an Attached Growth Sequential Batch Reactor. *International Journal of Environmental Research and Public Health.* 15 (8).

Banerjee, J. et al. 2016. Effect of Drying Methods and Extraction Time-Temperature Regime on Mango Kernel Lipids. *International Journal of Food Science and Nutrition* https://doi.org/10.15436/2377-0619.16.048.

Bengtsson, L. P. and J. H. Whitaker, eds. 1988. *Farm Structures in Tropical Climates—A Textbook for Design.* Rome: Food and Agriculture Organization and Swedish International Development Cooperation Agency. https://www.fao.org/3/S1250E/S1250E00.htm#Contents.

Bingo, M., M. Basitere, and S. Ntwampe. 2019. Poultry Slaughterhouse Wastewater Treatment Plant Design Advancements. 16th SOUTH AFRICA Int'l Conference on Agricultural, Chemical, Biological & Environmental Sciences (ACBES-19) 18–19 November 2019 Johannesburg (S.A.)

Bosak, V. et al. 2016. Performance of a Constructed Wetland and Pretreatment System Receiving Potato Farm Wash Water. *Water.* 8 (163). https://doi.org/10.3390/w8050183.

Bowser, T. and J. Nelson. 2021. Slaughterhouse Water Use and Wastewater Characteristics. Division of Agricultural Sciences and Natural Resources • Oklahoma State University Id: FAPC-240.

Bregnballe, J. 2015. A Guide to Recirculation Aquaculture: An Introduction to the New Environmentally Friendly and Highly Productive Closed Fish Farming System. In *FAO Eurofish Rep.* Rome: Food and Agriculture Organization of the United Nations (FAO) and EUROFISH.

Carawan, R. E. 1991. Processing Plant Waste Management Guidelines for Aquatic Fishery Products. *Sea Food and the Environment Pollution Prevention Short Course.* North Carolina State University.

Carawan, R. E., J. V. Chambers, and R. R. Zall. 1979. Seafood Water and Wastewater Management. North Carolina State University, The North Carolina, Agricultural Extension Service.

Cardenas, A. et al. 2009. Electrochemical Oxidation of Wastewaters from the Instant Coffee Industry Using a Dimensionally Stable RuIrCoOxanode. *ECS Trans.* 20 (1). pp. 291–299. https://doi.org/10.1149/1.32683 97.

Castine, S. A. et al. 2013. Wastewater Treatment for Land-Based Aquaculture: Improvement and Value-Adding Alternatives in Systems of Australia. *Aquaculture Environment Interaction.* 4. pp. 285–300.

Chandrasasi, D., R. Haribowo, and B. K. Wardana. 2021. Design for Wastewater Treatment Plants (WWTP) for Cattle in a Single House Model with a Bio Filter Process. *IOP Conference Series: Earth and Environmental Science.* 641. 012016.

China. 2016. Typical Characteristics of Municipal WWTP Intake Nationwide (North star network for water treatment, in Chinese). 全国典型城市污水处理厂进水水质特征-北极星水处理网. bjx.com.cn (accessed August 2023).

China Network for Wastewater Treatment. 2015. Cattle Husbandry Wastewater Treatment Design. 养牛废水处理方案. dowater.com.

China Research Academy on Environmental Sci (CRAES) et al. 2017. *Explanation for the Draft Update of Discharge Standard for Water Pollutants from Slaughtering and Meat Processing.* 屠宰与肉类加工水污染物排放标准(征求意见稿) -编制说明-2017. https://www.mee.gov.cn/gkml/hbb/bgth/201711/W020171115320356555089.pdf.

Ching, Y. C. and G. Redzwan. 2017. Biological Treatment of Fish Processing Saline Wastewater for Reuse as Liquid Fertilizer. *Sustainability.* 9 (7). p. 1062.

Coppola, D. et al. 2021. Fish Waste: From Problem to Valuable Resource. *Marine Drugs.* 19.116. https://www.ncbi.nlm.nih.gov/pmc/articles/PMC7923225/.

Cristóvão, R. O. et al. 2012. Chemical and Biological Treatment of Fish Canning Wastewaters. *International Journal of Bioscience, Biochemistry and Bioinformatics.* 2 (4).

Cristovao, R. O. et al. 2015. Fish Canning Industry Wastewater Treatment for Water Reuse—A Case Study. *Journal of Cleaner Production.* 87. pp. 603–612.

DAIRexnet. 2019. Water System Design Considerations for Modern Dairies—DAIReXNET. extension.org.

Daud, N. and C. S. Anijiofojor. 2017. Livestock Wastewater Generation and Farm Management: The Gap Analysis. *Acta Horticulturae.* 1152 (1). pp. 265–272.

El-Kamah, H. et al. 2010. Treatment of High Strength Wastewater from Fruit Juice Industry Using Integrated Anaerobic/Aerobic System. *Desalination.* 253. pp. 158–163.

Food and Agriculture Organization of the United Nations (FAO). 1992. Wastewater Characteristics and Effluent Quality Parameters. In *Wastewater Treatment and Use in Agriculture.* Rome.

FAO and World Health Organization. 2023. *Safety and Quality of Water Used in the Production and Processing of Fish and Fishery Products—Meeting Report.* https://doi.org/10.4060/cc4356en.

Ferraciolli, L. et al. 2017. *Potential for Reuse of Effluent from Fish-Processing Industries. Revista. Ambiente & Água.* 12 (5, Sep-Oct).

Fox, J. et al. 2014. Water Sprinkling Market Pigs in a Stationary Trailer. 1: Effects on Pig Behaviour, Gastrointestinal Tract Temperature and Trailer Micro-Climate. *Livestock Science.* 160.

Ghangrekar. Wastewater Discharge Standards for Food and Fruit Processing Industry. ghangrekar.com.

Giang, N. et al. 2021. *Recycling Wastewater in Intensive Swine Farms: Selected Case Studies in Vietnam.* Tokyo: Faculty of Agriculture, Kyushu University.

Gil, M.I. and A. Allende. 2018. Water and Wastewater Use in the Fresh Produce Industry: Food Safety and Environmental Implications. In F. Pérez-Rodríguez, P. Skandamis, and V. Valdramidis, eds. *Quantitative Methods for Food Safety and Quality in the Vegetable Industry. Food Microbiology and Food Safety.* Springer, Cham.

Gómez-Sanabria, A. et al. 2020. Sustainable Wastewater Management in Indonesia's Fish Processing Industry: Bringing Governance into Scenario Analysis. *Journal of Environmental Management.* 275 (1 December). 111241.

Government of Ireland. *National Development Plan—Enterprise Ireland: Transforming Irish Industry 2009 Sustainable Practices in Irish Beef Processing.* Dublin.

Holland, J. 2018. New Research Recovers Nutrients from Seafood Processing Water, Finds Multiple Recycling Opportunities. In *Seafood Source.*

Hussan, A. et al. 1992. Wastewater Treatment and Use in Agriculture. *FAO Irrigation and Drainage Paper.* No. 47.

Hussan, A. et al. 2019. Fish Culture Without Water!!! Does Aquaculture Contribute to Water Scarcity? Is Aquaculture Without Water Possible? *Int. J. Fish. Aquat. Stud.* 7. pp. 39–48.

International Finance Corporation (IFC). 2007a. Environmental Health and Safety Guidelines for Fish Processing. Washington, DC.

———. 2007b. Environmental Health and Safety Guidelines on Food and Beverage Processing. Washington, DC.

———. 2007c. Environmental Health and Safety Guidelines on Meat Processing. Washington, DC.

———. 2007d. GHS Guidelines for Mammalian Livestock Production. Washington, DC.

Ijanu, E.M. et al. 2020. Coffee Processing Wastewater Treatment: A Critical Review on Current Treatment Technologies with a Proposed Alternative. *Applied Water Science.* 10 (11). https://doi.org/10.1007/s13201-019-1091-9.

Islam, Md. S. et al. 2004. Waste Loading in Shrimp and Fish Processing Effluents: Potential Source of Hazards to the Coastal and Nearshore Environments. *Marine Pollution Bulletin.* 49. pp. 103–110

J-U-B ENGINEERS, Inc. 2009. *Fruit Juice & Puree Process Wastewater Pre-treatment and Land Application System.* ST0008077_SVZ_FinalEngRpt_2009-07-09.pdf.

Japan. *Part 3: Examples of Food Processing Wastewater Treatment.* https://www.env.go.jp/earth/coop/coop/document/male2_e/007.pdf.

Jemli, M. et al. 2015. Biological Treatment of Fish Processing Wastewater: A Case 2 Study from Sfax City (Southeastern Tunisia). *Journal of Environmental Sciences.* 30. pp. 102–112. https://doi.org/10.1016/j.jes.2014.11.002.

João, J. J. et al. 2020. Treatment of Swine Wastewater Using the Fenton Process with Ultrasound and Recycled Iron. *Revista Ambiente & Água.* 15 (3).

Keshem, A. H. M. et al. 2023. Microalgal Bioremediation of Brackish Aquaculture Wastewater. *Sci of the Total Environ.* 873.

Komlatsky, V. I. et al. 2022. The Importance of Water in Pig Farming. *BIO Web of Conferences.* 51. 03009. 10.1051/bioconf/20225103009.

Kurniasih, S. D. et al. 2018. Pollutants of Fish Processing Industry and Assessment of Its Waste Management by Wastewater Quality Standards. *E3S Web of Conferences.* 68. 03006.

Kurniawan, S. B. et al. 2021a. Aquaculture in Malaysia: Water-Related Environmental Challenges and Opportunities for Cleaner Production. *Environmental Technology and Innovation.* 24.

———. 2021b. Potential of Valuable Materials Recovery from Aquaculture Wastewater: An Introduction to Resource Reclamation. *Aquaculture Research.*

Lehto, M. et al. 2014. Water Consumption and Wastewater in Fresh-Cut Vegetable Production. *Agricultural and Food Science.* 23. pp. 246–256.

Lvfeng Environmental Engineering Company. Treatment of Slaughtering Wastewater. 其它肉类屠宰废水处理工程-肉类屠宰废水处理-河南绿丰环保工程有限公司. xxlfhb.com.

McCabe, B. et al. 2013. Assessing a New Approach to Covered Anaerobic Pond Design in the Treatment of Abattoir Wastewater. *Australian Journal of Multi-Disciplinary Engineering.* 10 (1). pp. 81–93.

Meat Research Corporation. 1995. Identification of Nutrient Sources, Reduction Opportunities and Treatment Options for Australian Abattoirs and Rendering Plants. North Sydney, Australia: Meat & Livestock Australia Limited.

Ministry of Ecology and Environment (MEE). 2009. Technical Specifications for Pollution Treatment of Livestock and Poultry Farms (HJ 497-2009). State Council of the People's Republic of China, Beijing.

———. 2010. Technical Specifications for Slaughterhouse and Meat Processing Wastewater Treatment Projects (HJ 2004-2010). 屠宰与肉类加工废水治理工程技术规范 (HJ 2004-2010). State Council of the People's Republic of China, Beijing.

———. 2020. Manual 135 on Pollution Co-efficient of Slaughtering and Meat Processing (Serial Manuals on Investigation and Cross-Checking Methods for Pollution Sources and Their Pollution Generation Coefficient). 排放源统计调查产排污核算方法和系数手册, 135 屠宰及肉类加工. State Council of the People's Republic of China, Beijing.

Misra, S.et al. 2020. Effect of Different Cleaning Procedures on Water Use and Bacterial Levels in Weaner Pig Pens. *PLoS ONE*. 15 (11). e0242495.

Mohsen, S. M. et al. 2012. Physical and Chemical Analysis of Wastes Generated During Processing of Mango and Orange Juices. *J. Food Industries & Nutr. Sci.* 2 (1). pp. 277–289.

Muthukumaran, S. and K. Baskaran. 2013. Organic and Nutrient Reduction in a Fish Processing Facility: A Case Study. *Intl Biodeterioration & Biodegradation.* 85.

Nagarajan, D. et al. 2019. Current Advances in Biological Swine Wastewater Treatment Using Microalgae-Based Processes. *Bioresource Technology*. 289. 121718.

Nagarajan, D. et al. 2019. Current advances in biological swine wastewater treatment using microalgae-based processes. *Bioresource Technology*. Volume 289, 121718. https://doi.org/10.1016/j.biortech.2019.121718.

North Dakota State University (NDSU). 2013. *Water Quality of Runoff from Beef Cattle Feedlots, Water Quality of Runoff from Beef Cattle Feedlots*. North Dakota. https://www.ag.ndsu.edu/publications/environment-natural-resources/water-quality-of-runoff-from-beef-cattle-feedlots/wq1667.pdf.

Novita, E. et al. 2022. Characterization of Laboratory Wastewater for Planning Wastewater Treatment Plant in University Campus in Indonesia. *Ecosystem Envirnmenta and Conservation*. Suppl. Issue, August. https://repository.unej.ac.id/handle/123456789/110616.

Nwuba, I. et al. 2019. Evaluation of the Characteristics of Wastewater from Slaughterhouses in Southeastern Nigeria for Design of Appropriate Treatment. https://www.researchgate.net/publication/330846223_Evaluation_of_the_characteristics_of_wastewater_from_slaughterhouses_in_South_Eastern_Nigeria_for_design_of_appropriate_treatment.

Othman, I. et al. 2013. Livestock Wastewater Treatment Using Aerobic Granular Sludge. *Bioresource Technology*. 133. 10.1016/j.biortech.2013.01.149.

Owino, W. O. and J. L. Ambuko. 2021. Mango Fruit Processing: Options for Small-Scale Processors in Developing Countries. *Agriculture*. 11. 1105.

Parihar, S. S. et al. 2019. Livestock Waste Management: A Review. *J. of Entomology and Zoology Studies.* 7 (3). pp. 384–393.

Park, J., J. H. Oh, and T. G. Ellis. 2012. Evaluation of an On-site Pilot Static Granular Bed Reactor (SGBR) for the Treatment of Slaughterhouse Wastewater. *Bioprocess Biosyst Eng.* 35 (3). pp. 459–468. doi.org/10.1007/s00449-011-0585-0.

Philippine Agricultural Engineering Standard (PAES). 2001. Agricultural Structures—Biogas Plant. In *PAES 413: 2001.* https://amtec.uplb.edu.ph/wp-content/uploads/2020/06/PNS-PAES-413-2001-Agricultural-Structures-Biogas-Plant.pdf.

Picos-Benítez, A. R. et al. 2019. Biogas Production from Saline Wastewater of the Evisceration Process of the Fish Processing Industry. *Journal of Water Process Engineering.* 32.

Pongthornpruek, S. 2017. Swine Farm Wastewater Treatment by Constructed Wetland Planted with Vetiver Grass. *Environment and Natural Resources Journal.* 15 (2). pp. 13–20. https://www.thaiscience.info/journals/Article/ENRJ/10985904.pdf.

Puchlik, M. 2016. Application of Constructed Wetlands for Treatment of Wastewater for Fruit and Vegetable Industry. *J. of Ecological Engineering.* 17 (1). pp. 131–135.

Puchlik, M. and J. Struk-Sokołowska. 2017. Comparison of the Composition of Wastewater from Fruit and Vegetables As Well As Dairy Industry. *E3S Web of Conferences.* 17. 00077. 10.1051/e3sconf/20171700077.

Quayson, J. and E. Awere. 2017. Water-Use and Conservation in the Commercial Vehicle Washing Industry in Urban Ghana: The Case of Cape Coast Metropolis. *IRA-International Journal of Technology & Engineering.* https://research-advances.org/index.php/IRAJTE/article/view/1088.

Rajakumar, R. et al. 2011. Treatment of Poultry Slaughterhouse Wastewater in Upflow Anaerobic Filter under Low Upflow Velocity. *International Journal of Environmental Science and Technology.* 8 (1). pp. 149–158.

Ribeiro, F. H. M. and L. P. Naval. 2017. Technologies for Wastewater Treatment from the Fish Processing Industry: Reuse Alternatives. *RBCIAMB.* 46. pp. 130–144. DOI: 10.5327/Z2176-947820170303.

Rodrigues, A. S. et al. 2022. Treatment of Fruit Processing Wastewater by Electrochemical and Activated Persulfate Processes: Toxicological and Energetic Evaluation. *Environmental Research.* 209 (June). 112868.

Salminen, E. 2002. Finnish Expert Report on Best Available Techniques in Slaughterhouses and Installations for the Disposal or Recycling of Animal Carcasses and Animal Waste. http://www.vyh.fi/eng/orginfo/publica/electro/fe539/fe539.htm.

Sannik, U. et al. 2015. Calculation Model for the Assessment of Animal By-product Resources in Estonian Meat Industry. *Agronomy Research*. 13. pp. 1053–1063.

Sarangi, K. et al. 2022. Sustainable Utilization of Pineapple Wastes for Production of Bioenergy, Biochemicals and Value-Added Products: A Review. *Bioresource Technology*. 351. 127085.

Shende, A. D. et al. 2022. Water Consumption, Wastewater Generation and Characterization of a Slaughterhouse for Resource Conservation and Recovery. *Water Practice and Technology*. 17 (1). pp. 366–377.

Suri, S. et al. 2022. Current Applications of Citrus Fruit Processing Waste: A Scientific Outlook. *Applied Food Research*. 2. 100050.

Tay, J. et al. 2022. Seafood Processing Wastewater Treatment. Nanyang Technological University, Singapore.

Tello, A., R. A. Corner, and T.C. Telfer. 2010. How Do Land-based Salmonid Farms Affect Stream Ecology? *Environmental Pollution*. 158. 1147–1158.

United Nations Environment Program (UNEP). 2000. Cleaner Production Assessment in Fish Processing. Nairobi. https://www.unep.org/resources/report/cleaner-production-assessment-fish-processing.

United States Environmental Protection Agency (USEPA). 2002. Development Document for the Proposed Effluent Limitations Guidelines and Standards for the Meat and Poultry Products Industry Point Source Category. Washington, DC.

Uttamangkabovorn, M. et al. 2005. Water Conservation in Canned Tuna (Pet Food) Plant in Thailand. *Journal of Cleaner Production*. 13. 547e555.

Valta, K. et al. 2017. Review and Assessment of Waste and Wastewater Treatment from Fruits and Vegetables Processing Industries in Greece. *Waste Biomass Valor*. 8. pp. 1629–1648.

VanDever, K. 2017. *Cooling Dairy Cattle in the Holding Pen - FSA4019 (uada.edu)*. University of Arkansas, United States Department of Agriculture, and County Governments Cooperating.

Vanotti, M. B. et al. 2014. *Generation 3 Treatment Technology for Diluted Swine Wastewater using High-Rate Solid-Liquid Separation and Nutrient Removal Processes*. https://www.ars.usda.gov/ARSUserFiles/60820500/Patents_presentations_papers/ASABE_paper_141901300.pdf.

Vásquez, Z. S. et al. 2019. Biotechnological Approaches for Cocoa Waste Management: A Review. *Waste Management*. 90 (May). pp. 72–83.

Venugopal, V. and A. Sasidharan. 2021. Seafood Industry Effluents: Environmental Hazards, Treatment and Resource Recovery. *Journal of Environmental Chemical Engineering*. 9. 104758.

WorldAtlas. What Is Coffee Wastewater? https://www.worldatlas.com/articles/what-is-coffee-wastewater.html.

Yeo, S. E., F. P. Binkowski, and J. E. Morris. 2004. *Aquaculture Effluents and Wastes By-products.* https://repository.library.noaa.gov/view/noaa/37672/noaa_37672_DS1.pdf.

Zachritz, W. H. and R. F. Malone. 1991. Wastewater Treatment: Options for Lousiana Seafood Processors. Lousiana Sea Grant College Program.

Zema, D. A. et al. 2019. Wastewater Management in Citrus Processing Industries: An Overview of Advantages and Limits. *Water.* 11. 2481; doi:10.3390/w11122481; www.mdpi.com/journal/water.

Ziara, R. M. M. et al. 2018. Characterization of Wastewater in Two US Cattle Slaughterhouses. *Water Environment Research.* 90 (9). pp. 851–863.

www.ingramcontent.com/pod-product-compliance
Lightning Source LLC
Chambersburg PA
CBHW050054220326
41599CB00045B/7410